Breeding Chickens For Egg Production
A Study of Annual and Total Egg Production in Chickens

by Utah Agricultural College Experiment Station

with an introduction by Jackson Chambers

This work contains material that was originally published in 1916.

This publication is within the Public Domain.

This edition is reprinted for educational purposes and in accordance with all applicable Federal Laws.

Introduction Copyright 2017 by Jackson Chambers

Self Reliance Books

Get more historic titles on animal and stock breeding, gardening and old fashioned skills by visiting us at:

http://selfreliancebooks.blogspot.com/

Introduction

I am pleased to present yet another title on Poultry.

The work is in the Public Domain and is re-printed here in accordance with Federal Laws.

As with all reprinted books of this age that are intended to perfectly reproduce the original edition, considerable pains and effort had to be undertaken to correct fading and sometimes outright damage to existing proofs of this title. At times, this task is quite monumental, requiring an almost total "rebuilding" of some pages from digital proofs of multiple copies. Despite this, imperfections still sometimes exist in the final proof and may detract from the visual appearance of the text.

I hope you enjoy reading this book as much as I enjoyed making it available to readers again.

Jackson Chambers

BREEDING FOR EGG PRODUCTION—PART I
A Study of Annual and Total Production.
BY E. D. BALL, BYRON ALDER and A. D. EGBERT.

INTRODUCTION

The average annual egg production of different breeds of fowls, the production that may be expected under different environmental conditions, the average length of life, the total probable production of an individual and the distribution of this production through the life cycle, are all subjects of vital importance to every poultryman. They are also subjects upon which there must be reliable information before accurate breeding work can be carried on. A search through the literature shows that information on all these subjects is extremely meagre.

Much interest in annual egg production has been aroused in recent years by egg laying contests. These contests have been an important factor in establishing standards of maximum egg production and in determining the relative value of the different breeds for this purpose. These records have, however, shown the maximum production possible to obtain from a few rigidly selected specimens carefully chosen for exceptional vigor and maturity as well as for supposed inherent laying qualities, but give only slight indication of the average production of the flocks from which they come.

When the breeding work with White Leghorns was taken up at the Utah Experiment Station in 1907, it was taken for granted that the first year's production of a hen was the highest, and that it was a reliable indication of her inherited productivity. Both of these points were at that time accepted as unquestioned and it was planned to use these as the foundation of this experiment. Fortunately, however, as the yards were not filled at the time, nearly all birds were kept and the second and third years' records obtained. As has been pointed out in a previous publication,[1]* over three fourths of the second year's records and many of the third in this flock were higher than the first, and there was little correlation between the first year's record and those of the later years.

---

*All references are given in a Bibliography at the back.

After studying these records carefully it seemed necessary, before intelligent selection could be practiced to discover:

First, a reliable measure of the productivity of a hen.

Second, the average productive life of a hen (that should be considered in selection). With these two points established it was thought that it would then be possible to plan an experiment to determine the value of selection in improving a strain of fowls.

Utah Station Bulletin No. 135 gives the records of the flocks up to October 31, 1913, and discusses annual and total production. The present paper gives two more years' records of these flocks and the records of two additional flocks and likewise confines itself to a discussion of the annual and total production, other factors being reserved for later publication.

HISTORICAL

The early literature on egg production is like that in other scientific lines—filled with loose and vague statements. Results are given without mention of details that would seriously modify their value. In many early experiments the number of individuals used was too small to give any reliable data. Most of the early work concerned itself with cost and methods of production rather than production itself, and the annual production is only given, if given at all, as the measure of some other factor.

It was also found necessary to reject all references to popular or semipopular literature on account of the amazing unreliability of these sources.

EXPERIMENTAL RESULTS

Wheeler[40] of the Geneva, N. Y. Experiment Station in 1891 gives results of feeding four pens of second year hens of different breeds. The record was very low, averaging 53 eggs, which was said to be "but little lower than the first year." Again in 1896 he[38] gives results of feeding experiments on twenty-three two-year-old Leghorns in two pens with egg records of 77 and 93, or an average for the second year of 85 eggs. The second pen's record was higher than its first year record. Sixteen Cochins two years old gave 64 and 48 eggs. This was a "pronounced decrease" from the first years records.

Gilbert[6] of the Canadian Experimental Farm, (1895 to 1906) gives part year* records as follows, the number of months given being the first of the laying year.

Year	'95	'96	'97	'98	'99	'00	'01	'02	'03	
Number of months	10	12	7	8	7	6½	7	7	7	Avg.
White Leghorns	121	70	99	68	76	83	52	69	79	80
B. Plymouth Rocks	113	73	90	71	88	51	69	57	73	76
Black Minorca	98	75	99	49	57	76	71	68	57	72

The number of hens of each breed was small, averaging about ten. He also gives the total number of each flock and its production. These flocks average over two hundred fowls, of which about one-third were pullets at first, increasing to nearly two-thirds towards the last without materially changing the flock average, which was very low, averaging only fifty-nine eggs.

Year	'94	'95	'96	'97	'98	'99	'00	'01	'02	'03	Avg
No. pullets		93	53	53	63	80	96	136	123	72	85
No. hens		125	151	151	157	125	106	118	100	161	132
Total No. of flock	185	218	204	204	220	205	202	249	223	233	214
Average Product	48	44	56	70	63	62	64	48	63	70	59

In the 1899 Report[6] a test of age is given as follows, about ten hens of each breed and age or seventy-six in all, being used.

	Pullets	Year old	Old hens
Leghorns	76	56	45
B. Plymouth Rocks	88	51	56
Black Minorca	57		70
Average Production	74	54	57

In the 1904 Report[6] he states: "Records of egg-laying by pullets and hens in our department, extending over eight years, go to show that pullets which laid well during their first winter did not make as good records the next, when hens. It was also shown that pullets which were poor layers during their first winter season did remarkable well as hens the next one."

He[6] also gives in 1905-06 report the first year's records of

---

*All through this bulletin, the laying year from November 1st. to October 31, is designated by the numerals of the year in which the greater part falls e. g. '95 refers to year Nov. 1. 1894, to Oct. 31, 1895.

several breeds, the flocks ranging from twelve to twenty-three each.

Year	'05	'06
White Leghorn	81	77
B. Plymouth Rocks	66	66
White Wyandottes	63	

Besides these records the following records of older fowls were given.

Year	'05	'06	'07	'08	'09
	3 Yr. Old	2 Yr. Old			2 Yr. Old
B. Plymouth Rock	62	76			83
White Wyandottes		75			97

Gowell of the Maine Station gives, 1900-03, by far the best and most complete data on egg production thus far published including the complete records of large flocks of three varieties. He gives[7] monthly records of 39 hens for the year 1899. These were only those hens of the flock that laid over 160[1] or under 100 eggs, the ones between being omitted, so that the flock averages for that year can only be approximated. Assuming that the average of the medium layers would be 130 eggs—the average of the extremes—and we get the averages given below for that year. The next three flocks give 139, 132 and 133, or an average of 135 for this same group, so that this figure is probably low and each average given should probably be raised from three to five eggs.

In 1903[9] he gives the monthly record of all of the next three flocks of each breed. Omitting those that died during the year, the flock averages are as follows:

Breed	No. of fowls	'99	'00	'01	'02	'03	'04	'05	'06	'07	Avg.
B. Rocks	50-126	[127]	136	143	156	136*	118	135	140	114	139
W. Wyandottes	33-72	[128]	130	133	132						131
L. Brahmas	16-53	[129]	107								118

He gives[8] the individual first and second year records of the high layers that were kept for breeding work and also gives

---

1 In this experiment a full year from time of laying was taken. The eggs laid after Nov. 1st '99 have been subtracted in giving the average, which lowers the record of some of these hens below 160.

*Pearl and Surface[22] give the annual averages obtained in the next five years of Gowell's work.

a few third year records. This was the first time individual records extending over more than one year had been published. These birds were of course selected after one year's production and are not representative of the flock as a whole, but only the upper third or fourth.

Barred Rocks

Flock of	No. of hens	'99	'00	'01	'02	1st. Yr.	2nd Yr.	3rd. Yr.
1898	27	157	115	100		157	115	100
1899	4		205	111	79	205	111	79
1900	26		(High)	188	104	188	104	
1900			(Low)	78	46			
					Average	183	110	90

White Wyandottees

| 1899 | 21 | 150 | 106 | | | 150 | 106 | |

Light Brahmas

| 1898 | 14 | 133 | 74 | | | 133 | 74 | |

These records must be compared with those of egg laying contests or other highly selected flocks. The "low" records of 1900 are of course the other extreme and are omitted from the average.

Dryden[2], 1897 to 1900, gave results of the Utah Station on feeding tests of R. C. B. Leghorns, from which the following table has been compiled. The records were from very small numbers, usually four or five in a pen and from three to five pens per year. The high first year records show that the birds used were selected individuals for this work from much larger flocks as in egg laying contests.

Flock hatched	1897	1898	1899	Years of laying.		
				1st.	2nd.	3rd.
1896	157	133	117	157	133	117
1897		161	132	161	132	
1898			144	144		
			Average	154	133	117

He gives records of four pens of B. P. Rocks, which averaged 136 eggs in 1899.

Dryden[3] in 1905 gives records of two year tests of individuals of different breeds, which are summarized as follows:

	First year records Lowest	Highest	First year Average	2nd. Year
13 R. C. B. Leghorns	170	228	193	157
4 S. C. W. Leghorns	161	203	183	95
3 W. P. Rocks	174	209	192	129
5 B. P. Rocks	122	212	154	110
16 W. Wyandottes	115	216	170	111
41 Average			178	125

These were all selected as "high layers" after the first year's records were made, and the number of individuals in the original flocks or the flock average from which these were selected, was not given nor even the years in which these records were made. They may have been made in different years, as the records reported cover the period from November 1896 to November 1902.

Some of these individuals were kept through the third year with averages as follows:

	1st.	2nd.	3rd.	4th.
4 R. C. B. Leghorns	199	188	119	60
7 W. Wyandottes	174	116	71	67

One hen from each breed was kept through the fourth year with records of 60 and 67 respectively.

Stewart and Atwood[32] of West Virginia Station 1899 tested the effect of age on egg production. Three pens containing three and four-year-old hens laid more eggs than three pens of pullets. The experiment was run two hundred and ten days.

In 1906 they give comparison of W. Leghorns and Mongrels in 1905 as follows:

	1905
50 W. Leghorns average	117
50 Mongrels average	96

In 1908 they[32] give another record as follows:

	1907
600 W. Leghorns average	116

Rice[30] of Cornell. 1907-08, gives results of feeding on egg production and in 1912 gives results of selecting strong and weak looking birds in 1908. The favorable conditions are tabulated.

Year	'07	'08
White Leghorns (pullets)	113	126
White Leghorns (2 yr. old)	101	109
White Leghorns (3 yr. old)	90	

Linfield[18] of Montana, 1906-07, gives flock average as follows:

Year	No. fowls	'05	'06
Barred P. Rock pullets	8-12	80	68
Barred P. Rock hens (1 yr.)	10		71
White Leghorn pullets	(pen)	126	
White Wyandotte pullets	8		70
White Wyandotte hens	8		70

Lewis[15] of New Jersey, gives the following flock records:
    19 White Leghorns in 1906 average 122
   160 White Leghorns in 1911 average 111
   79 R. I. Reds in 1913 average 99
   80 B. P. Rocks in 1913 average 80

In 1905-06, Nelson[19a] of New Jersey gives results of two pens of ten birds each (Barred and White Rocks) with an without male for two years as follow:

	1905 Jan. 1st. to Oct. 31	1906 (2nd. Yr.) Nov. 1st. to Oct. 31
Pen 1—Rocks (with male)	83	126
Pen 2—Rocks (without male)	82	118

Lewis in 1913[15] gives results of feeding high and low protein to pens of fifty fowls. The high protein ration gave an average of 140 eggs.

He also gives average per cent production of different breeds, the flocks varying from eighty to six hundred and sixty-two fowls. The production translated into eggs is as follows:

    White Leghorns...... 118
    R. I. Reds............ 107
    Barred Rocks........ 100

Jacobs[11] of Arkansas, 1908, gives records of small flocks as follows, and notes that the Wyandottes were late maturing and did not lay until January:

	No. of fowls	1907
B. P. Rocks	10	141
S. L. Wyandottes	6	110
W. Leghorns	7	109

The Irish Department of Agriculture[10] obtained records for the year 1908 on one hundred and twenty-five flocks representing over five thousand fowls. These gave an average production of 120 eggs. Some breed records are as follows:

	1908
White Leghorns	135
Buff Orpingtons	133
Buff Wyandottes	129
Barred P. Rocks	109
Avg. all flocks	120

Opperman and Waite[21] of Maryland, 1911, tested the effect of age on egg production, and decided that third year production is obtained at a loss. Their figures do not bear this out, however. They started with a flock of more than two hundred and forty hens and at the end of the first year selected the sixty highest layers and kept them through the second and third years with results as follows:

		1st. Yr.	2nd. Yr.	3rd. Yr.
240	Total flock	127	114	
60	Selected high layers	171	149	115
180	Remainder of flock	112	103	

It will be noticed that the third year record of the selected flock was 115, while the second year of the total flock was only 114 and the first year record of the one hundred and eighty left after taking out the sixty selected ones was still less, being 112.

Nixon[20], 1910, gives a correlation study of the three year record of eighty-eight White Leghorn hens. These hens were used in experimental work and were subject to various tests (not given) and therefore only general conclusions can be drawn from the records which follow:

	1st. Yr.	2nd. Yr.	3rd. Yr.	3 Yr. Avg.
88 White Leghorn hens	92	97	86	91

These records are low; it will be noticed, however, that they are almost alike, the second year being the highest, and the three year average is only one egg less than the first year.

Lloyd and Elser[17], 1911, give results of a survey of poultry conditions in Ohio as follows:

	Avg. eggs.
18 farm flocks of an average of 121 hens gave	71
12 town flocks of an average of 46 hens gave	70
1 commercial flock of an average of 333 hens gave	141
Total 31 flocks. Total 3063 hens.	

Sherwood and Buss[31] of Ohio, 1913, carried on two tests of the cost of egg production with the following egg yields:

	1911	1912
214 White Leghorns	120	
313 White Leghorns		128
151 B. P. Rocks		114

Rice[28], 1913, of Cornell Station, discusses the distribution of egg production and gives three year records of seventy-six hens as follows:

	No. of hens	1909 1st. Yr.	1910 2nd. Yr.	1911 3rd. Yr.	3 Yr. Average
Flock A	38	154	135	123	137
Flock B	38	142	124	109	125
Average	76	148	130	116	131

The records of flock A are high for all three years. Any one of these alone would be quite usual, but the combination gives a high three year average. The other flock is very close to the average throughout.

Ball, Turpin, and Alder[1], 1914, of the Utah Station give records of six flocks of White Leghorns in which all hens were kept as long as they lived, the oldest flock having records for six years, the next for five, and so on down to the pullet year of the last flock. As these records are used in the summaries of this bulletin, they are not repeated here.

Rice[29], 1915, gives the average records for one hundred and sixty-nine White Leghorns for three years. These records were taken from thirty-eight hatched in 1909, sixty-three in 1910 and sixty-eight in 1911.*

	No. fowls	1st. Yr.	2nd. Yr.	3rd. Yr.	3 Yr. Avg.
Avg. 3 flocks of White Leghorns	169	137	124	109	123

The three year average of these flocks is practically the same as we have secured. The first year average is a trifle higher and the later years correspondingly lower than our records.

Philips[25] of Indiana, 1915, carried on feeding experiments with White Leghorn pullets for four years. He had twenty-five pullets per pen and the two pens each year that were well fed, averaged as follows:

	1911	1912	1913	1914	Avg.
25—Meat or fish scrap	125	145	142	114	132
25—Skimmilk	124	143	137	138	136
50—Average	125	144	140	126	134

RESULTS OF EGG LAYING CONTESTS.

The egg laying contests have been of immense value to the poultry industry, especially in giving popularity to the utility side of the subject. They have also been of great value in establishing standards of production and it is to be hoped that these established records will soon take the place of the mythical performances still so common in the popular press.

In considering these records it must be constantly borne in mind that these are the records of a few of the very best producers from the best flocks of the world. The entries are highly selected not only as to breeding strains but more especially for maturity and immediate performance. To illustrate how close this selection is, Kirkpatrick and Card[14] note that the production of the third contest at Storrs was lower than the previous ones and ascribe it in part to the change from five birds per pen to ten per pen so that "not quite such good specimens were selected for the pen."

The Hawkesbury Agricultural College of Richmond N. S.

---

*Distribution of flocks from private correspondence.

ANNUAL EGG PRODUCTION

Wales, appears to have held the first full year contest in 1902-03, the Utility Poultry Club of England having started winter egg laying contests in 1897-98 not extending to year contests until ten years later. The accompanying table shows the results of the Australian contests, as given by Thompson[36]. The White Leghorns being the most important breed, the first part of the table is devoted to them and is practically self-explanatory. The average first year (pullet) laying of the entire number of White Leghorns entered is shown in the first figures of the lowest diagonal series and down the first column at the right. The number of pullets competing each year is shown on the left.

	02-3	03-4	04-5	05-6	06-7	07-8	08-9	09-10	10-11	11-12	12-13	13-14	14-15	1st	2nd	3rd
								210	154	106						
									209	148	134	Three				
White Leghorns										203	162	123	year			
				201	145		Two				207	152	134			
24	136				195	138	year							136		
72		166				202	140	records						166		
96			166				197	135						166		
120				168				187	158					168		
138					175	(145)				206	141			175	145	
138						195	(138)				201	147		195	138	
114							199	(140)	(106)			193	163	199	140	106
204								193	(135)	(134)				193	135	134
114									185	(158)	(123)			185	158	123
192										192	(141)	(134)		192	141	134
168											190	(147)		190	147	
168												187	(163)	187	163	
216													191	191		
								Average of all 1st year						185		
							Average of those held for 2 years							198	146	
						Average of those held for 3 years								207	154	124
Black Orpingtons																
	143	168	160	158	178	169	177	172	150	187	175	168	178	168		
Silver Wyandottes																
	143	162	145	168	171	160	171	155	156	157	152	135	167	158		

Results of the Hawkesbury A. C. Egg Laying Contests.

Beginning with the flocks of 1906-07, the highest layers were kept to make second year records. Their first year records were of course higher than the average of the flock for the Leghorns and are shown in the first figures of the middle diagonal series followed by the second year's record each in its respective laying year spaces. Beginning with the pullets of 1908-09 that were making their second year record in 1909-10, some of the second year hens were kept a third year. Only the best of these were selected so both the first and second year records are above the average of the totals made. The records for the Leghorns are shown in the upper set of diagonals. The second and third year averages of all records made are also shown at the right. The other two breeds are given in a later table. It must be remembered that the first year records are of a restricted group, selected before laying began; that the second year's records are of a selected group of these, the selection made on first year records and, that the third year records are of a still further selected group. Therefore these records cannot be compared with the three year records of entire flocks, such for example as shown in the Utah and Cornell Station results. Instead they are only to be compared with the highest ten of these flocks and even here they are not strictly comparable.

Besides the White Leghorns, which made up nearly two-thirds of the pens in the later contests, only two other breeds have run through the series. These have had an average of about fifty hens each and their records are included for comparison with the general purpose breeds in the American contests.

TWO AND THREE YEAR RECORDS OF THE GENERAL PURPOSE BREEDS HAWKESBURY CONTESTS.

	Second Year Laying				Third Year Laying					
of	B. Orpingtons		S. Wyandottes		B. Orpingtons			S. Wyandottes		
	1st. Yr.	2nd. Yr.	1st. Yr.	2nd. Yr.	1st. Yr.	2nd. Yr.	3rd. Yr.	1st. Yr.	2nd. Yr.	3rd. Yr.
07-08	188	116	180	129	......	......	......	......	......	......
08-09	172	111	174	123	......	......	......	......	......	......
09-10	192	137	174	143	......	......	......	......	......	......
10-11	183	127	206	88	210	154	106	......	......	......
11-12	159	105	177	122	209	148	134	......	......	......
12-13	191	124	189	112	165	123	99	180	147	92
13-14	179	120	176	120	196	136	110	188	112	106
14-15	177	123	178	156	......	......	......	......	......	......
Avg.	180	120	182	124	195	140	112	184	130	99

Egg laying contests were begun in **South Australia** at Magill in 1903-04 and continued, except for the year 1906-07, at Roseworthy until 1912-13, then at Parfield from 1913-14 on. Laurie[16] gives detailed results from which the following has been compiled.

ANNUAL EGG PRODUCTION

Year	03-4	04-5	05-6	06-7	07-8	08-9	09-10	10-11	11-12	12-13	13-14	14-15	Avg.
W. Leghorn						208	199	205	181	191	184	196	195
B. Orpington						178	176	175	161	151	155	138	162
S. L. Wyandottes						177	183	168	152	148	143	160	162
Average of all fowls entered	132	117	171		180	190	186	192	176	182	178	187	184
Average of three breeds						188	186	183	165	163	161	165	

There were from 400 to 800 fowls entered each year of which sixty to eighty per cent were White Leghorns. This great preponderance of Leghorns brings the average of all fowls considerably above the average of the three breeds.

Egg-laying competitions were begun in England by the Utility Poultry Club[37] in 1897 and have been continued to date. For the first ten years the tests were confined to the winter production, but since October 1907 annual tests have also been carried on. Only two years of these records have been obtained. The Harper Adams Agricultural College started a competition in 1912-13 which has been continued to date. The records of the five leading breeds are given in the accompanying table.

	Utility Club 07-8	Utility Club 13-14	Harper Adams 13-14
White Leghorns	117	202	201
White Wyandottes	149	183	188
White Orpingtons		164	152
Buff Rocks	136	189	194
Rhode Island Reds		195	161
Average	134	187	179

Murphy[19] gives (1912-15) the records of the first three Irish contests from which the records of the three leading breeds have been tabulated. The White Wyandottes were the leading breed, followed by the Rhode Island Reds with the Leghorns third in numbers of those enumerated. The average production of the three breeds was in the same order, the Wyandottes leading with a three-year average of 145, the Reds following with 143, while the Leghorns averaged in the three years only 137 eggs. Miss Murphy writes that the records of the fourth contest will be much higher throughout. These contests started October first and lasted only eleven months.

	Irish Contests.		
	First 12-13	Second 13-14	Third 14-15
White Leghorn	111	139	161
White Wyandotte	129	140	167
Rhode Island Red	153	141	134
Average of 3 Breeds	131	140	154

Murphy also gives the records of second and third year laying of selected high layers from the contest birds as follows:

	Irish Contests		
Number Hens	First Year	Second Year	Third Year
12	184	133	......
22	159	118	......
7	181	155	105
Average	175	135	105

The records in the above table were received too late to be inserted in Table No. VI but they would not materially change the averages given.

A number of egg laying contests were started in America in 1912 (Nov. 1911) and have continued to date. Two others were started later and are still running. The four breeds of fowls that were the most popular, judged by the number of entries, and which were represented in all contests, have been tabulated below. These breeds were represented in each contest by a number of individuals ranging from thirty to over four hundred, except in three cases where only a single pen was entered. These three records are bracketed and have not been used in the averages. The largest number of entries in every case has been of White Leghorns.

	Storrs, Conn.					The National at Mt. Grove, Mo.					The Missouri at Mt. Grove, Mo.	
	1st. 1912	2nd. 1913	3rd. 1914	4th. 1915	Avg.	1st. 1912	2nd. 1913	3rd. 1914	4th. 1915	Avg.	1st. 1914	2nd. 1915
W. Leghorns	163	171	156	158	162	143	173	172	136	156	159	160
W. Wyandotte	161	165	153	148	157	125	151	190	128	149	124	[166]
R. I. Red	159	162	136	152	152	156	136	171	146	152	140	126
B. Rock	154	137	140	147	145	126	148	176	135	146	151	[209]
Av. of breeds	159	159	146	151	154	138	152	177	136	151	144	143

ANNUAL EGG PRODUCTION   17

	Vancouver, B. C.					Philadelphia North America Newark, Del.		Avg. of all
	1st. 1912	2nd. 1913	3rd. 1914	4th. 1915	Avg.	1st. 1914	2nd. 1915	
W. Leghorns	......	166	172	169	......	178	179	164
W. Wyandottes	117	153	181	165	154	195	180	156
R. I. Reds	132	152	134	168	147	167	167	153
B. Rocks	116	135	162	[223]	138	160	147	145
Avg. of breeds	......	152	162	167	152	175	168	153

The records of the egg laying contests given above were taken from the reports of these contests (12), (13), (14), (27), (34), (35) except for 1915 which was obtained from Purvis, Breeders Gazette (26).

EXPERIMENTAL RESULTS AND CONCLUSIONS.

General Considerations.

In a previous bulletin of the authors[1] which is still available, the general details of methods and equipment have been given, together with the record of the first four flocks to that date. The present publication gives two more years' records of each of the four flocks together with four years' records of the 1911 flock and three years' records of the flock of 1912, making fifteen yearly flock records added to the data available, all of which has been summarized in the tables which follow.

Number of Hens Completing Records.

Table No. I gives the number of hens that completed each years record in each flock to November 1st, 1915. It was intended to start each flock with about one hundred pullets. Variation in success in hatching and brooding caused this number to vary, but the average has been a little above that figure.

TABLE NO. I—NUMBER OF HENS COMPLETING FULL YEARS RECORDS BY FLOCKS.

Flock of	No. of Hens completing Records for the Year.								No. of Hens completing Records for the Year of Laying.							
	'08	'09	'10	'11	'12	'13	'14	'15	1st.	2nd.	3rd.	4th.	5th.	6th.	7th.	8th.
'07	121	91	58	31	24	14	6	4	121	91	58	31	24	14	6	4
'08		135	98	49	34	23	13	10	135	98	49	34	23	13	10	
'09			96	73	48	31	25	14	96	73	48	31	25	14		
'10				121	100	37	26	18	121	100	37	26	18			
'11					85	53	39	33	85	53	39	33				
'12						164	137	100	164	137	100					
'13							115	85	115	85						
'14								70	70							
Total Record									897	649	331	155	90	41	16	4
Average No. in Flock									112	93	55	31	23	14	8	4
Grand Total									2183							

From this table, it is seen that eight flocks ranging in age from nine years down to one year have been studied, including thirty-six yearly flock records, embracing 2183 complete yearly individual records by days.

Method of Handling Flocks.

Before taking up the analysis of these records, it is well to have the following points in mind:

That these records are all from descendants of a single flock of fowls, no outside blood having been introduced since the work began.

That care has been exercised to prevent close inbreeding, separate strains having been constantly maintained.

That they have always been fed on a commercial basis, neither force feeding for egg production nor to force moulting, having been practiced.

That with few exceptions they have been handled in flocks of ten or twenty with a rooster during the breeding season.

That practically all pullets that completed their first year's record, have been kept as long as they lived or are alive today. A few have been stolen and a few from the earlier flocks were culled out to make room.

That these flocks are practically unselected as far as production is concerned, no basis for selection having been found in the earlier years, and since that time breeding has been carried on from high, medium, and low producers.

That each pullet flock has been rigidly selected for vigor and the later ones are from strains showing good fertility, hatchability, chick survival, and length of life.

A Study of the Flock Records.

In all the calculations, the first year record includes only eggs laid between November 1st and October 31st, of the following year; eggs laid before November 1st have been recorded, but not included in these figures. In figuring averages of individual records they are given to the nearest whole number. Where the average is halfway between two numbers, the higher one was always taken, as this will not make up for the number of eggs lost where trap nest records are used nor for those stolen, so that the numbers given will still be too low for the actual production.

ANNUAL EGG PRODUCTION

To simplify the reading of the tables, the year record from November 1st to October 31st is designated by the numerals of the calendar year in which the greater part of the record falls.

The First Year Production of a Flock.

The production of the pullet year of a flock depends upon a combination of many factors, some of which are difficult or impossible to control. All flocks of all ages are subject to the effect of the environment during the time the record is made, but the pullet year record is affected equally or even more seriously by the environment of the previous season. Under normal influences, chicks hatched at a given time and brooded in a given way will be ready to begin laying early in November. Under our conditions, April to early May hatched chicks with proper brooding are ready for their work at the proper time. Earlier hatching will often result in early laying and at least a partial moult in November and December, while later hatched chicks that do not get started before cold weather often do not start laying until late winter or early spring. In this way one unfavorable season may affect two year's records—first the pullets of that year directly, and secondly, the pullets of the next season by giving them a poor start.

Table No. II gives the average production of all the individuals in each flock that completed the given years record, as shown in Table No. I, while Table No. III gives the average production of all the hens that finished the third year of laying, as given in Tables 14-17, Bulletin 135 and Table IX and X of this bulletin.

TABLE NO. II—AVERAGE YEARLY PRODUCTION OF FLOCKS (INCLUDING ALL HENS COMPLETING ONE OR MORE YEARS RECORDS.)

	Average of Flocks for Year.								Average of Flocks for the Year of Laying.								3 Yr. Avg.
	'08	'09	'10	'11	'12	'13	'14	'15	1st.	2nd.	3rd.	4th.	5th.	6th.	7th.	8th.	
'07	107	135	117	78	87	93	44	52	107	135	117	78	87	93	44	52	120
'08	......	136	105	82	90	89	58	58	136	105	82	90	89	58	58	......	108
'09	......	......	155	101	113	103	65	47	155	101	113	103	65	47	......	......	123
'10	......	......	......	86	137	136	76	53	86	137	136	76	53	......	......	......	120
'11	......	......	......	......	114	133	104	75	114	133	104	75	......	......	......	......	117
'12	......	......	......	......	......	152	97	85	152	97	85	......	......	......	......	......	111
'13	......	......	......	......	......	......	116	126	116	126	......	......	......	......	......	......	......
'14	......	......	......	......	......	......	......	63	63	......	......	......	......	......	......	......	......
'15	......	......	......	......	......	......	......	......	......	......	......	......	......	......	......	......	......
			Average (omitting 1914)						124	119	106	84	74	66	51	52	116

This latter table is the more valuable for comparative purposes, as the first three years records are made by the same individuals.

In comparing the two tables it will be seen that there is an average difference between the first year records of eight eggs. This increase in Table III is due to two causes: natural selection eliminating by death more of the weaker individuals and therefore poorer layers, and the culling out in some of the flocks, of a few of the poorer ones as "left overs" when the pens were made up.

The average first year production as shown by Table II omitting the abnormally low record of 1915, which was caused by the combination of a poor season and the installation of a new system of brooding—which was not ready in time—and we have 124 as the average of all hens. Taking only those that completed three years gives us a first year average of 132 as shown in Table III.

TABLE NO. III—AVERAGE YEARLY PRODUCTION OF FLOCKS (INCLUDING HENS COMPLETING THREE OR MORE YEARS RECORDS.)

	Average of Flock for the Year.								Average of Flock for Year of Laying.							3 Yr. Avg.
	'08	'09	'10	'11	'12	'13	'14	'15	1st.	2nd.	3rd.	4th.	5th.	6th.	7th.	8th.	
'07	117	146	117	78	87	93	44	52	117	146	117	78	87	93	44	52	127
'08		153	117	82	90	89	58	58	153	117	82	90	89	58	58		117
'09			156	111	113	103	65	47	156	111	113	103	65	47			127
'10				94	151	136	76	53	94	151	136	76	53				127
'11					119	139	104	75	119	139	104	75					121
'12						153	101	85	153	101	85						113
Average									132	128	106	84	74	66	51	52	122
Average of highest year									150								

Even these figures do not represent the maximum laying capacity of these flocks because half of the flocks have made their highest records in their second year. Taking the average of the highest year of each flock from Table II would give an average of 139 eggs and from Table III an average of 150 eggs for the highest year of each flock. In this way most of the effects of unfavorable environment can be eliminated from the measure of production.

Comparing these results with all the available results obtained with sufficient numbers of individuals and where there was apparently no great amount of selection at the beginning, and we get Table No. IV with variations between seventy-seven and one hundred and fifty-six and a rough average of about one hundred and twenty-eight, which agrees very well with our work. Figure 1 gives a graphic representation of the most important of these records and compares them with the records of selected birds in egg laying contests.

TABLE NO. IV—FIRST YEAR RECORDS OF APPARENTLY UNSELECTED FLOCKS.																	
	'99	'00	'01	'02	'03	'04	'05	'06	'07	'08	'09	'10	'11	'12	'13	'14	Avg.
Maine	(127)	136	143	156	136	118	135	140	114*								134 BPR
Maine	128	130	133	132				125		126	148(137)						131 WW
Cornell							125	113	126	148(137)							130 WL
N.J							122				111		118				117 WL
Ohio												120	128				124 WL
Ind												125	144	140	126		134 WL
Canada								81	77								(79) WL
Misc. (Irish 135 1908) (Md. 112 1908) (Nixon 92) (117 W. V.)																	
Average (Can. and Misc. out)																	128

*No yearly records have been published since that of 1907 of the Maine flock.

The results given for all ages in the Canadian experiments and those of the Montana Station, are so much lower than the others that they are not used in the averages.

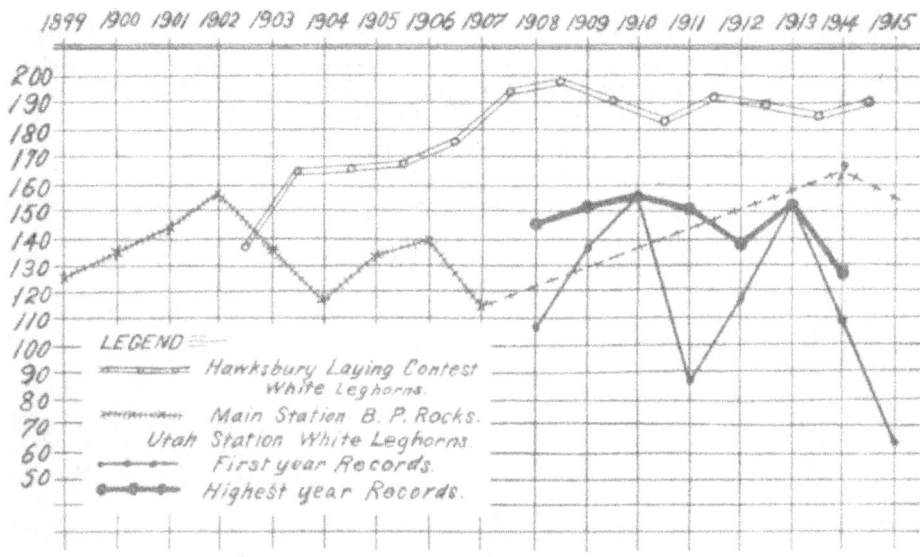

Fig. 1.

Taking up the pullet records of egg laying contests, and other tests embracing very small numbers which were evidently selected in the same way, we get Table No. V which shows the maximum production possible to obtain where all the individuals used are selected for maturity and are supposed to be in laying condition at the beginning of the year. Only a small per cent of the flock is ever represented in such tests so the averages show maximums and can only be compared if at all with the averages of highest tens of flocks. The results of some experiments of Dryden's are included here on account of the early date at which they were carried on.

The White Leghorns are given for each contest and the highest average producer of the general purpose breeds for each region.

The striking thing about Table V is the very high records made in the Australian contests, especially in the later years when the numbers of Leghorns have run up to five hundred and six hundred birds. The average of the American contests is not very high when it is considered that the Maine Station made an unselected flock average of 156 in 1902 and an eight-year average of 134 with Barred Rocks, while the Utah Station has made an unselected flock average of 156 and a six-year average of 150 with White Leghorns.

A comparison of these records with the records of the highest first year tens of each of our six flocks, shows that the selection in the American flocks has not been very high. The highest tens

TABLE NO. V—FIRST YEAR RECORDS OF SELECTED INDIVIDUALS FROM LARGER FLOCKS.

		'02	'03	'04	'05	'06	'07	'08	'09	'10	'11	'12	'13	'14	'15	Avg.
Hawks.	WL	136	166	166	168	175	195	199	193	185	192	190	187	191		185
Hawks.	BO	143	168	160	158	178	169	177	172	150	187	175	168	178		168
Rosew.	WL							208	199	205	181	191	184	196		195
Rosew.	BO							178	176	175	161	151	155	138		162
Eng.	WL							117					202			
Eng.	WW							149					183			
Storrs.	WL											162	171	156	158	162
Storrs.	WW											161	165	153	148	157
Mo.	WL											143	173	172	136	156
Mo.	WW											125	151	190	128	149
Van C.	WL												166	172	169	(169)
Van C.	WW											117	153	181	165	154
Dryden	BL	157	161	141	(from '97-'99)											(154)
Dryden	BR		136													(136)

of our flocks represent an average of about the highest fifth and their record of 176 is the same as the average of American and Australian contests and much above the American contest average. The selection in the American contests must be much less than the upper fifth if we grant that their flock averages are the same as ours and the average of the different records at hand as shown in Table No. IV is very close. If instead of taking the first year records, the highest tens of the highest years of each flock were taken, the record would be as follows:

Utah's highest tens' highest year 187-202-200-197-184-193=194 which is above the Austrailian average. It is probable that the high Australian average is partly the result of weeding out and building up flocks during the years of contests, and partly the result of a more equitable climate that allows of a more even distribution of laying throughout the year.

Another thing brought out in Table V is that the maximum production of the White Leghorn is on the average, considerably higher than that of the general purpose breeds. The highest producing one of these breeds has been selected for comparison in each case and yet in every contest except those in Great Britian the Leghorn has averaged higher than the general purpose breeds. There is a much greater difference in the Austrailian contests than in the American.

It is to be regretted that there are not more annual records of unselected flocks of the general purpose breeds, so that comparison could be made with the Leghorns. The few scattered records that have been published indicate that there is the same difference as in the contest records.

To summarize the study of first year production, it appears that the average for White Leghorn flocks under normal conditions of feed and care is from 123 to 128 eggs; that the average American maximum production is about 162 and the Australian 190 eggs; and that the heavier breeds fall somewhat below this amount, the greatest variation occurring in Australia.

The Second Year Production of a Flock.

Table Nos. II and III give the second year production of seven flocks. The average of all hens in Table No. II being 119, while those that finished the third year as shown in Table No. III averaged 127 eggs in their second year. These averages are only four or five eggs below the corresponding first year averages of the same flocks [124-132.] and indicate that there is little difference between first and second year production under our conditions. The variation in flock averages is not nearly as great in the second year as it was in the first, the records ranging between 100 and 145 eggs, with the average falling very close to the medium point between these figures.

Very little experimental work has been done on egg production beyond the first year, and still less where an entire unselected flock has been kept for later production, as has been the case in these experiments. Nearly all the work except that at Cornell and Maryland has been with the extremely high producers of the first year, only "high producers" in most cases those with first year records above 160, being kept. Table No. VI brings together the records of production after the first year and shows the years in which the American records were made.

TABLE NO. VI—FIRST, SECOND AND THIRD YEAR RECORDS OF VARIOUS FLOCKS.

	'97	'98	'99	'00	'01	'02	'03	'04	'05	'06	'07	'08	'09	'10	'11	'12	Avg 1st	Avg 2nd	Avg 3rd
Dryden; Br. L.	157	133	117														159	133	117
Br. Leghorn		161	132																
Gowell; B. P. R.			157	115	100	79													
				205	111	104													
					188														
					(78	46)	omit selection low										183	110	90
Gowell; W. W.				150	106														
Dryden; Br. L.					193	157	119	60											
W. W.					170	111	71	67									182	134	95-64
Mont. B. P. R.								(80	71)	omit									
Md. W. L.			240									127	114						
Md. W. L. selected			60									171	149	115			127	114	[115]
Nixon; W. L. (Yr.?)	88											92	97	86					
Rice; W. Leg	76											148	130	116			180	146	119
W. L. (Yr. ?)	169											137	124	109			126	117	104
Hawkesbury 8 yrs. tests W. L.																	198	146	124
Hawkesbury 4 yrs. tests W. L.																	[207	154]	112
Hawkesbury 8 yrs. tests B. O.																	180	120	
Hawkesbury 4 yrs. tests B. O.																	[195	140]	99
Hawkesbury 8 yrs. tests S. W.																	182	124	
Hawkesbury 2 yrs. tests S. W.																	[184	130]	110
Irish Egg laying test																	184	133	
All nonselected Leghorns																	127	116	110
All selected Leghorns																	180	146	119
All selected General Purpose																	177	116	93
Utah highest tens Leghorns																	176	146	118

The average of the nonselected flocks of 127 eggs first year and 116 eggs second year does not differ materially from the Utah averages, the first year being four eggs higher and the second year three lower than our results, or a difference of eleven eggs between the two years, while ours shows only four or five.

The nonselected flocks of Leghorns show a slightly lower second year record than our flocks, while the selected Leghorns show a much higher second year average than ours. These selected flocks of Leghorns and the highest tens of our flocks show the same second year average (146 eggs) and very slight difference in the other years.

There were no records of flocks of nonselected general purpose breeds available but the selected ones of these breeds gave a much lower (30 eggs) second year record than that of the Leghorns, and the same falling-off continued into the third year.

The Third Year Production of a Flock.

Table No. III gives the third year production of the six different flocks, including 331 hens with a range from 82 to 136 and an average of 106, which is a little below the median figure indicating that 136 was an extreme record.

Table No. VI gives the average third year record of nonselected Leghorn flocks as 110, of selected Leghorns as 119. The general purpose breeds averaged 93 eggs in the third year and only the better producing breed in each test were included.

The two averages of non-selected flocks of White Leghorns, 106 and 110, are so close as to indicate a very definite potential third year laying capacity for this breed.

The Three Year Average as a Measure of Production.

The right side of Table No. III shows the years of production of the different flocks arranged in columns with the three year average along the right margin. A glance will show that there is no agreement between the first year record and the three year average of a flock. In fact in Table No. VII in which the first year's record of the flocks are arranged in order, the lowest first, it is seen that the three flocks with the low first year records gave higher three year averages than those of the three high flocks.

The low first years averaged 110, while the high ones

averaged 154, their corresponding three year averages were 125 and 119, the hens in the low flocks producing 18 eggs apiece more in three years than their sisters in the high flocks. It will be noticed that with one exception the second year records are in the reverse order from the first year ones. Figure 2 brings this out graphically. If the first two years' records are added and arranged in order in the same way, it will be found that the third year records of the three low ones average 108, while the three high ones average only 104.

The two flocks with the lowest three year averages finished their three years' records in the two most unfavorable years, 1911 and 1915, and thus did not get a chance to make up the deficiency. Flocks that passed the 1911 year as either a first or second year, made later records high enough to balance the low one.

In general it appears that there is no relation between the first year production of a flock and the total production in three years; that a high first year production is always followed by a low second year record and vice versa; that the second year record alone does not influence the third year, but that the total production of the first two years slightly influences inversely the third year production, giving a fairly equal three year average.

It appears from these studies that our flocks of White Leghorns have a fairly definite potential laying capacity that finds expression in three years of production regardless of whether the first year's record is high or low.

It appears further that no greater uniformity is reached in succeeding years and that, therefore, the three year average is the most accurate measure of productivity of a flock as well as the first accurate one that can be applied under our conditions.

Turning to selected flocks in which the first years's average is always far above, and only a slight indication of the real average of the flock from which they were selected, and Gowell's two flocks of Barred Rocks are the only Station flocks available. Here the flock with the lowest first year record averaged 25 eggs per hen more in the next two years than did the high first year flock. The Hawkesbury contests contain records of 8 flocks of each of the breeds, that were kept for 2 years. Dividing each of these breeds into the four highest and four lowest first year records and the White Leghorns and Silver Wyandottes show higher average second year records of six eggs each for the low flocks, while the Black Orpingtons show the reverse, with an eleven egg

TABLE NO. VII—FLOCKS OF HENS WITH THREE YEARS RECORD ARRANGED IN ORDER OF FIRST YEAR PRODUCTION.

First Year Record	94	117	119	153	153	156
Second Year Record	151	146	139	117	101	111
Third Year Record	136	117	104	82	85	113
Second &Third Year Record	287	263	243	199	186	224
Three Year Average	127	127	121	117	113	127
Avg. of First Year Record		110			154	
Avg. of the Three Year Product		125			119	

difference. The three year records of the same breeds, though few in number, show the same tendencies as the two year records.

Production After the Third Year.

After the third year the production gradually declines. In the White Leghorns an apparent fairly regular decline occurs at the rate of about 10 eggs per year, the number of records in the later years being too few to give anything more than an approxi-

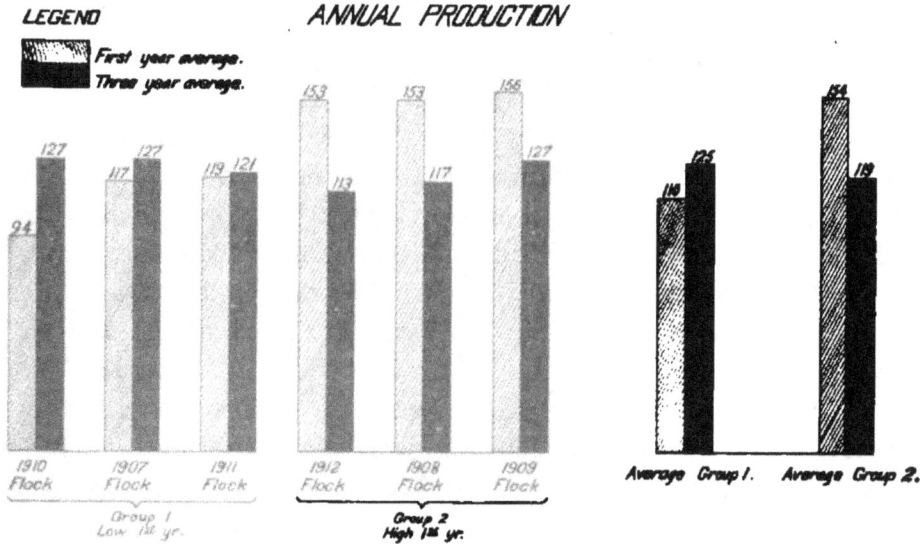

Fig. 2—Note the influence of high or low first record on the three year average of a flock.

mation. With the few records at hand it does not seem to be as much affected by previous production as by seasonal influence.

The first three flocks have finished six years' records. The last three of the six years average 86, 79 and 72 for the three flocks in the order named. There appears to be no relation between these and the averages of the first three years, but they do vary inversely with the first year production, although the variation is so small that it may not be significant.

Environmental Influence on Egg Production.

A study of the left half of Table No. III will show that the seasons have had a marked influence on egg production. The year 1908 was probably an unfavorable one, as the 1907 flock made a much higher record the second year than it did the first, and even the third year equalled the first one.

Unfavorable Years.

The year 1911 was a notoriously bad one in the poultry industry of the Intermountain region. More hens died this year than during any other year on record. Eggs failed to hatch and chicks were so weak that they died in large numbers in the brooders, so that the 1911 flock was small in numbers and late hatched, and still later in maturing and remained low in vigor and vitality. This resulted in a poor record for this flock in 1912 and only fair ones following this, ending with a poor three year record. Not only did this year affect the eggs and chicks but it reduced the egg yield of every flock in the plant to a lower level than they had made previously or than they made in the year following, despite the increased age. This influence was not confined to the Station poultry plant, but was widely distributed throughout the Intermountain region. It even appears to have been general. The 1911 records in Table No. IV are all low, suggesting its influence in Indiana, Ohio and New Jersey, and the 1910-11 records of the Hawkesbury Australian contest are the lowest in an eight year series. These records cover the winter and spring of 1911, which was apparently the time of trouble.

Again in 1914 we see a dropping off of all records from the oldest flock to the youngest followed by a slightly greater decline in 1915, indicating a generally unfavorable condition through two seasons. The unfavorable conditions of the 1915 season were again general throughout the Intermountain region; calls for chicks continuing six weeks later than usual because of generally poor results in incubation and brooding.

Favorable Seasons.

The years 1911 and 1915 were undoubtedly unfavorable, but whether the records in 1914 were low because of a definitely unfavorable season or on account of an inevitable reaction from the high production of the two favorable years 1912 and 1913, is not so certain. The low first year record of the 1913 flock indicates that it was not a favorable year, but the 110 record is an indication that it was not as bad as 1911 when the first year record was 86.

The years 1909 and 1910 were favorable as evidenced by the first year records and the high second year of the 1907 flock, while 1912 and 1913 were even more favorable, all records being high except the first year record of the 1911 flock, which as explained before was late in maturing and low in vitality. The extremely low record made by the flock of 1914 cannot be charged against the 1915 season nor even the season of 1914 when they were hatched, as a new brooder house was built in 1914 and it was not in proper shape in time for that year's hatch, so that this flock was in poor shape to enter the 1915 season. On this account this record has been omitted in the averages.

Figure 3 shows graphically how the years influence egg production. The lower curve shows the calculated seasonal in-

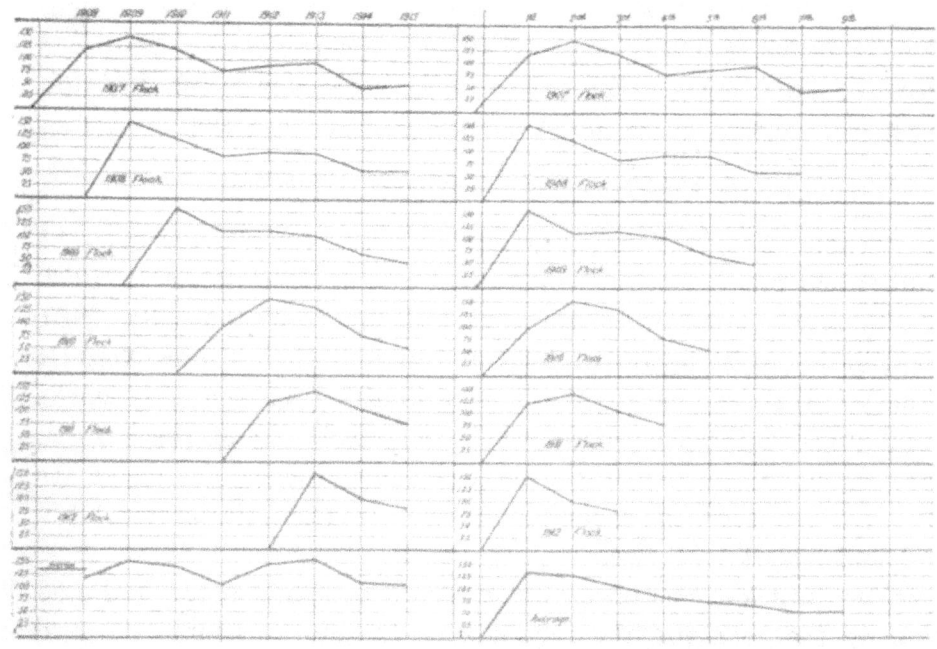

Fig. 3—Showing the influence of the season on production.   Fig. 4—Showing the influence of age on production.

fluence on average first year production. The flock curves show how this influence affected the flocks of different ages. The influence of the two unfavorable periods is evident on each flock producing at that time. The definite drop in 1911 shows markedly in every case.

The influence of the season was determined in two ways: first, by determining the deviation of each record from the average of that age, and taking the average deviation of the records made in a given year as a factor; second by taking the average of the first, second and third year records made in each year and then determining the deviation from the average of these averages, correction being made for the late maturity of the 1911 and 1914 flocks.

Figure 4 shows how age influences production. The lower curve represents the average production to date and is as near a representation of normal production in the White Leghorn as the present data will furnish. Selected flocks giving higher first and second years records would probably decline more rapidly with age. The different flocks are arranged to show the modification of this curve as influenced by the environment they experienced.

The curve of production is based on the three years' records and is drawn that way so that the flock records can be compared directly with it. If the entire flock were considered, the first two years would be 124 and 119 instead of 132 and 128.

A Study of the Individual Records.

Tables Nos. XIV to XVIII of Utah Bulletin No. 135 give the individual records of the first four flocks up to October 31, 1913.

The continuation of these records up to October 31, 1915, will be found in Table No. VIII of this bulletin. Tables Nos. IX and X give the records of all hens in the 1911 and 1912 flocks that finished three or more year's records. All of these flocks are arranged in order on the three year average. The flocks are divided into tens, the incomplete tens in the middle so that highest and lowest tens, twenties, etc. can be compared. In all these tables a bracket around a yearly record means that the hen died during the year and the record was, therefore, incomplete, and a bracket around a total production, except in Table VII, means that the last incomplete year has been included. In Table No. VIII a bracket around a total production means that the hen was still living October 31, 1915, so that the total is incomplete. Table

ANNUAL EGG PRODUCTION

TABLE NO. VIII—ADDITIONAL PRODUCTION OF FIRST FOUR FLOCKS* BY YEARS WITH TOTAL PRODUCTION TO DATE.

Flock of 1907			Flock of 1908			Flock of 1909			Flock of 1910		
'14	'15	Total	'14	'15	Total	'14	'15	Total	'14	'15	Total
......	......	616	[1]	......	625	129	97	[913]	[36]	......	602
......	......	505	......	......	498	......	......	598	35	......	520
......	......	771	......	......	611	......	......	564	144	[49]	672
......	......	537	......	......	562	......	......	550	147	136	[742]
......	......	562	73	110	[858]	47	17	[653]	[9]	......	464
......	......	470	......	......	549	84	29	766	47	[0]	488
......	......	471	[47]	......	741	108	[0]	721	[0]	......	435
......	......	470	[48]	......	755	......	......	469	38	4	[475]
......	......	602	1	15	[684]	66	1	692	127	80	632
......	......	520	37	60	[698]	......	......	455	111	70	[605]
......	......	504	90	[0]	753	[0]	......	593	[0]	......	414
......	......	458	......	......	538	......	......	444	[19]	......	430
[37]	......	775	99	[0]	773	65	[48]	699	15	0	[419]
......	......	465	54	52	[734]	......	......	587	95	77	[573]
......	......	569	115	97	[856]	......	......	434	96	39	[535]
26	18	761	92	90	[778]	47	[24]	605	104	70	[573]
......	......	589	......	......	399	......	......	554	46	6	[446]
......	......	553	......	......	393	134	85	[748]	[27]	......	411
[17]	......	657	[31]	......	640	51	[0]	536	[0]	......	381
......	......	617	[0]	......	574	34	2	[485]	[29]	......	405
61	106	[901]	......	......	392	45	......	556	......	......	......
......	......	605	[0]	......	574	111	......	603	......	......	......
......	......	408	72	82	[643]	61	18	[554]	......	......	......
......	......	486	......	......	418	99	63	[605]	......	......	......
......	......	402	......	......	361	......	......	443	......	......	......
......	......	483	......	[0]	385	[32]	......	518	......	......	......
37	......	726	......	......	341	79	......	553	......	......	......
[0]	......	650	......	......	567	......	......	401	......	......	......
[27]	......	700	......	......	434	......	......	......	......	......	......
......	......	572	......	......	......	......	......	......	......	......	......
......	......	635	......	......	......	......	......	......	......	......	......
......	......	377	......	......	......	......	......	......	......	......	......
......	......	428	......	......	......	......	......	......	......	......	......
50	73	730	......	......	......	......	......	......	......	......	......
......	......	430	......	......	......	......	......	......	......	......	......
......	......	380	......	......	......	......	......	......	......	......	......
[22]	......	763	......	......	......	......	......	......	......	......	......
......	......	392	......	......	......	......	......	......	......	......	......
......	......	469	......	......	327	54	48	[570]	......	......	......
[7]	......	656	[33]	......	585	[41]	......	502	......	......	......
......	......	530	......	......	449	61	52	[568]	......	......	......
86	[12]	811	......	......	393	67	97	[578]	140	75	[588]
......	......	363	......	......	313	......	......	422	2	1	[372]
......	......	337	......	......	311	......	......	343	117	105	[591]
......	......	323	56	51	[594]	67	......	522	75	......	433
......	......	311	......	......	306	24	[0]	427	57	51	[464]
......	......	319	......	......	538	......	......	354	78	......	432
......	......	382	......	......	351	91	74	[576]	94	64	[512]

*Tables No. 14 to 18 of Bulletin No. 135 give the records of these flocks previous to 1914. Only the total production is given of those that died before the end of 1914 laying season. The individuals are arranged in tens as in the previous tables. The smaller flocks being separated in the middle to facilitate comparison of the highest and lowest tens, etc. A bracket [ ] around a yearly record indicates that the hen died before the close of the year. A bracket around a total record indicates that the

(Table No. VIII—Continued)

Flock of 1907			Flock of 1908			Flock of 1909			Flock of 1910		
'14	'15	Total	'14	'15	Total	'14	'15	Total	'14	'15	Total
.....	.....	300	.....	.....	376	.....	.....	336	27	24	[389]
.....	.....	320	[0]	.....	362	.....	.....	305	69	43	437
.....	.....	355	.....	.....	269	13	[0]	388	73	[0]	386
.....	.....	402	.....	.....	252	18	.....	402	62	[5]	380
.....	.....	275	.....	.....	234	[47]	.....	385	[0]	.....	305
.....	.....	326	.....	.....	221	.....	.....	268	[19]	.....	319
5	10	[348]	.....	.....	233	.....	.....	230	52	[0]	336
.....	.....	249	[37]	.....	369	.....	38	[372]	[30]	.....	309
.....	.....	293	12	5	[248]	42	32	[366]	62	70	[407]
.....	.....	238	52	22	[276]	27	[2]	277	76	31	[365]
44	52	503	58	58	494	65	47	.....	76	53	.....

TABLE NO. IX—PRODUCTION BY YEARS OF ALL HENS IN FLOCK OF 1911 THAT FINISHED THREE OR MORE YEARS RECORDS (ARRANGED ACCORDING TO THREE YEAR AVERAGE) 1911 YEAR RECORD.

Hen No.	'12	'13	'14	'15	Totals 3 Yr.	4 Yr.
174	181	205	176	95	562	657
151	158	191	162	161	511	672
300	137	178	191	63	506	569
254	108	193	180	145	481	626
173	155	141	147	144	443	587
152	171	131	130	144	432	576
257	165	156	102	82	423	505
193	145	119	159	(37)	423	(460)
179	133	174	112	52	419	471
261	146	153	111	148	410	558
286	85	189	135	124	409	533
285	186	172	50	1	408	409
169	128	172	103	(34)	403	(437)
284	115	147	139	91	401	492
159	142	160	94	118	396	514
255	127	168	97	129	392	521
256	162	188	42	1	392	393
147	151	132	107	104	390	494
182	50	164	157	46	371	417
160	144	121	105	(3)	370	(373)
251	146	146	71	2	363	365
192	118	143	83	36	344	380
195	78	137	129	70	344	414
293	104	136	103	113	343	456
283	122	129	92	53	343	396
264	99	106	127	90	332	422
260	157	102	68	.....	327	.....
301	46	145	117	71	308	379
198	110	109	80	47	299	346
157	119	109	68	(10)	296	(306)
280	80	113	99	11	292	303
262	56	134	85	(38)	275	(313)
266	69	128	77	77	274	351
154	120	96	49	29	265	294
292	80	109	67	1	256	257
200	55	105	95	84	255	339
171	105	55	88	53	248	301
178	89	83	40	44	212	256
252	81	82	15	45	178	223
Average	119	139	104	75		

TABLE NO. X—PRODUCTION BY YEARS OF ALL HENS IN THE FLOCK OF 1912 THAT FINISHED THREE OR MORE YEARS RECORDS. (ARRANGED ACCORDING TO THREE YEAR AVERAGE).

Hen No.	'13	'14	'15	Average 3 Yr.	Total 3 Yr.
363	195	187	144	175	526
312	175	151	190	172	516
344	223	152	136	170	511
353	178	168	155	167	501
337	193	159	129	160	481
315	200	147	133	160	480
417	182	190	106	159	478
348	152	178	137	156	467
326	193	123	142	153	458
364	193	95	167	152	455
346	194	125	136	152	455
399	162	121	165	149	448
339	163	152	128	148	443
332	169	157	114	147	440
386	187	127	123	146	437
485	188	100	144	144	432
343	187	124	121	144	432
336	150	146	133	143	429
314	160	143	118	140	421
425	152	144	123	140	419
365	201	92	127	140	420
413	154	179	82	138	415
362	178	128	109	138	415
524	161	135	108	135	404
382	177	128	95	133	400
395	182	105	111	133	398
437	159	97	136	131	392
354	175	156	52	128	383
464	156	60	160	125	376
359	188	93	95	125	376
419	154	121	100	125	375
377	166	84	124	125	374
496	178	92	101	124	371
321	162	95	111	123	368
410	154	86	125	122	365
373	173	99	91	121	363
482	168	117	78	121	363
504	156	110	93	120	359
536	172	72	114	119	358
452	133	122	101	119	356
421	144	98	111	118	353
476	167	118	67	117	352
325	150	93	109	117	352
501	185	79	87	117	351
479	169	79	103	117	351
350	163	108	79	117	350
368	153	136	59	116	348
338	174	99	73	115	346
420	151	91	103	115	345
418	184	112	49	115	345
477	180	101	63	115	344
546	159	101	83	114	343

(Table No. X—Continued)

Hen No.	'13	'14	'15	Average 3 Yr.	Total 3 Yr.
472	197	95	47	113	339
537	133	110	94	112	337
462	126	117	87	110	330
369	130	114	84	109	328
469	126	113	78	106	317
424	153	119	42	105	314
545	165	72	77	105	314
497	122	118	73	104	313
530	152	89	68	103	309
375	148	78	81	102	307
330	119	87	96	101	302
503	159	82	58	100	299
459	156	89	54	100	299
429	130	108	61	100	299
367	138	70	91	100	299
539	106	101	92	100	299
543	148	100	46	98	294
366	210	78	2	97	290
500	171	82	36	96	289
463	118	102	66	95	286
422	92	95	98	95	285
532	142	102	37	94	281
486	159	61	61	94	281
460	175	26	78	93	279
435	126	71	81	93	278
541	156	82	37	92	275
534	113	111	50	91	274
441	106	94	70	90	270
525	141	86	42	90	269
428	196	70	0	89	266
493	146	44	75	88	265
433	120	87	57	88	264
439	112	88	63	88	263
465	114	78	70	87	262
423	111	93	57	87	261
430	145	104	10	86	259
440	106	66	79	84	251
522	142	92	13	82	247
457	121	73	53	82	247
436	128	64	50	81	242
432	81	103	44	76	228
547	102	62	61	75	225
529	115	51	22	63	188
361	126	39	9	58	174
471	77	60	28	55	165
328	146	1	0	49	147
518	68	32	45	48	145
515	114	0	1	38	115
Average	153	101	85	113	339

No. XI gives the records of all hens in the 1911 and 1912 flocks that finished only one or two years of laying.

TABLE NO. XI—PRODUCTION BY YEARS OF ALL HENS IN 1911 AND 1912 FLOCKS THAT FINISHED ONLY ONE OR TWO YEARS RECORDS.

Flock of 1911					Flock of 1912				
Two Years Records			One Year Record		Two Years Records			One Year Record	
Hen No.	'12	'13	Hen No.	'12	Hen No.	'13	'14	Hen No.	'13
294	81	99	177	135	327	173	98	316	167
196	105	71	298	49	331	75	1	318	205
279	156	138	297	91	335	126	52	322	133
185	128	174	290	31	340	208	97	324	191
176	109	122	287	86	357	179	26	345	168
155	185	125	273	128	371	186	118	349	197
282	51	116	272	89	391	188	73	352	219
275	127	150	265	67	392	189	154	360	88
278	109	96	263	72	393	143	128	374	131
191	115	116	259	105	394	248	158	378	179
288	106	102	194	50	401	145	125	381	116
153	187	112	190	111	406	145	106	533	116
303	110	65	189	132	415	143	100	384	123
183	167	50	186	142	416	153	173	385	110
			184	81	453	172	126	531	94
			175	53	467	133	88	403	56
			170	144	470	74	115	404	161
			168	123	480	144	82	405	147
			165	117	483	179	74	408	118
			156	125	542	155	78	409	139
			150	124	489	163	92	414	190
			589	64	517	125	68	426	133
			181	76	347	136	5	461	195
			172	163	356	77	26	498	152
			146	50	379	169	127	508	171
			253	44	320	159	57	511	70
			269	30	333	88	7	313	168
			276	61	334	167	145		
			258	115	372	143	59		
			363	84	376	150	89		
			591	105	487	165	77		
			180	37	505	197	97		
			(32)		528	144	110		
					323	159	10		
					427	186	41		
					351	132	122		
					412	138	10		

Distribution of Individuals in the Flocks.

The flock averages have been observed to vary between wide ranges from year to year and the question naturally arises whether the individual records all vary in the same way or whether the difference is due to a few high producing individuals in the high years and a corresponding number of low producers in the low years. Table No. XII gives the distribution of the

TABLE NO. XII—RELATIVE DISTRIBUTION OF HENS IN FLOCKS EACH YEAR ACCORDING TO PRODUCTION.

Flocks Years	220-239	200-219	180-199	160-179	140-159	120-139	100-119	80-99	60-79	40-59	20-39	0-19	Flock Avg.	No. Hens	Standard Deviation
1907															
1st. year			3	8	11	24	19	29	16	8	3		107	121	36.3±1.6
2nd. year		2	3	13	28	17	14	7	5	1	1		135	91	34.1±1.7
3rd. year			1	2	9	22	10	9	1	2	1	1	117	58	32.6±2.0
4th. year							9	8	6	6	2		78	31	25.3±2.2
5th. year						2		9	4	3	3	3	87	24	33.8±3.3
6th. year					1	1	4	6		1		1	93	14	32.2±4.1
7th. year								1	1	1	2	1	44	6	26.9±5.2
8th. year							1		1			2	52	4	42.4±10.1
1908															
1st. year		6	7	16	35	29	24	11	4	3			136	135	34.3±1.4
2nd. year		1	1	3	13	19	25	14	6	9	5	2	105	98	39.6±1.9
3rd. year					1	7	7	12	9	6	5	2	82	49	34.5±2.4
4th. year				1		7	10	5		4	5	2	90	34	41.5±3.4
5th. year					4		7	6	2	1	1	2	89	23	34.9±3.5
6th. year								1	3	2	3	1	58	13	32.7±4.3
7th. year							1	3	1	2	1	2	58	10	33.8±5.1
1909															
1st. year	3	7	18	14	18	22	5	7	2				155	96	37.2±1.9
2nd. year		1	3	3	3	14	10	18	10	4	6	1	101	73	43.4±2.4
3rd. year				1	7	17	9	6	4	2	2		113	48	32.3±2.2
4th. year				1	5	4	6	7	5	3			103	31	32.9±2.9
5th. year						2	2	3	7	6	3	2	65	25	32.0±3.1
6th. year								3	2	2	3	4	47	14	30.4±3.9
1910															
1st. year				1	3	12	21	34	26	19	4	1	86	121	29.4±1.3
2nd. year	1	2	11	16	21	18	14	7	6	3	1		137	100	39.8±1.9
3rd. year		1	1	6	11	8	5	4	1				136	37	30.2±2.4
4th. year					3	1	3	3	7	4	3	2	76	26	39.9±3.7
5th. year						1	1	1	6	2	3	4	53	18	34.4±3.9
1911															
1st. year			4	4	13	15	21	11	6	9	2		114	85	38.4±1.9
2nd. year		1	4	9	8	11	11	6	2	1			133	53	35.1±2.3
3rd. year			2	2	3	5	8	9	5	4		1	104	39	39.7±3.0
4th. year				1	4	2	3	5	4	7	2	5	75	33	46.7±3.9
1912															
1st. year	2	8	25	35	39	24	15	5	6	1			153	160	34.8±1.3
2nd. year			2	4	12	16	28	35	22	7	5	6	97	137	37.8±1.5
3rd. year			1	3	4	14	15	16	19	16	5	7	85	100	40.8±1.9
1913															
1st. year	1	2	4	8	14	23	22	19	19	1	3		115	116	39.1±1.7
2nd. year	2	3	5	18	16	16	12	9	5	3	2	6	126	97	51.2±2.5
1914															
1st. year							2	4	14	17	18	14	1	63	70	26.7±1.5

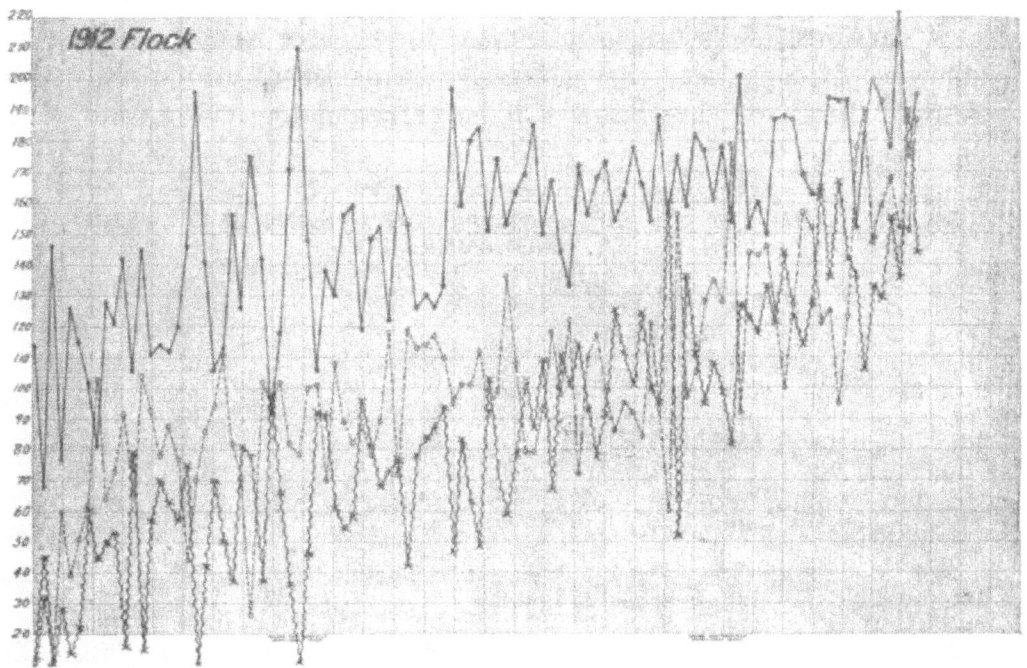

Fig. 5—A high first year average. Hens are ranked on three year average. First, second and third years record of a hen are on the same vertical line.

individual records by twenties throughout all years of all flocks and shows that the distribution varies with the variation in flock averages. In a high year the distribution occupies the upper spaces, leaving two or three vacant at the lower end, while in low years the reverse is true. The records of the flock of 1913 are an apparent exception to this rule, as they occupy nearly the entire series in both years. In the first year the removal of one or two individuals would correct this, but in the second year the distribution is extremely wide and accentuated at the lower end where it should not be. It is probable that the unfavorable 1915 season influenced this distribution. The effect of age in reducing numbers, lowering production, and scattering distribution is well shown in the older flocks.

Comparison of First, Second and Third Year Production.

Figures 1 and 2 in Bulletin 135 show graphically the wide variation in production of the individuals in high and low first year flock, and Figure 5 of this bulletin shows this same variation for another flock with high first year production and a low three year average. The individuals are arranged according to total production, the lowest first. A glance at these charts will show that this total production is made up of three widely separated records in the majority of cases. If the record of one year is high, the next year record will be correspondingly low, and vice versa.

TABLE NO. XIII—YEARLY EGG PRODUCTION ARRANGED BY TENS ON THREE YEAR AVERAGES.

	'08	'09	'10	'11	'12	'13	'14	'15	Avg. 3 Yr.	Total 3 Yr.
1907										
1st. 10	147	177	147	78	108	103	......	......	157	471
2nd. 10	134	156	134	94	84	74	26	18	141	425
3rd. 10	121	149	129	78	87	92	49	106	133	400
4th. 8	104	146	122	83	118	123	50	73	123	372
5th. 10	105	140	91	73	75	115	86	......	112	336
6th. 10	84	107	80	37	63	13	5	10	88	271
Flock Avg.	117	146	117	78	87	93	44	52	126	379
1908										
1st. 10	......	191	149	126	116	84	37	62	156	467
2nd. 10	......	174	134	97	113	108	90	80	135	405
3rd. 9	......	155	126	75	83	60	72	82	119	359
4th. 10	......	128	107	76	86	112	56	51	104	311
5th. 10	......	115	69	36	39	68	32	14	73	220
Flock Avg.	......	153	117	82	90	89	58	58	117	352
1909										
1st. 10	......	......	192	164	143	135	87	57	166	498
2nd. 10	......	......	179	120	131	115	66	44	143	429
3rd. 8	......	......	162	106	116	95	90	40	128	384
4th. 10	......	......	143	102	102	87	59	68	116	348
5th. 10	......	......	105	64	74	87	27	35	81	243
Flock Avg.	......	......	156	111	113	103	65	47	126	380
1910										
1st. 10				119	182	159	93	70	154	460
2nd. 10				90	164	142	71	48	132	396
3rd. 7				93	127	142	82	59	121	362
4th. 10				74	122	103	60	42	100	299
Flock Avg.				94	151	136	76	53	127	381
1911										
1st. 10					150	164	147	115	154	461
2nd. 10					129	161	103	77	131	393
3rd. 9					109	128	97	60	111	334
4th. 10					85	104	68	43	85	255
Flock Avg.					119	139	104	75	121	361
1912										
1st. 20						180	144	137	154	461
2nd. 20						167	109	106	127	382
3rd. 20						157	104	78	113	339
4th. 20						141	85	63	97	290
5th. 20						121	65	39	75	225
Flock Avg.						153	101	85	113	339

It will be observed that while flock averages vary from year to year, the records of the individuals making up these flocks vary even more widely in the same year, as well as following the variation of the flock as a whole.

Table No. XIII brings together the average of the tens[1] making up the six flocks that have finished three years' records arranged according to total production in three years.

A study of the table will show that the variation is evenly distributed through the different tens and the different years, indicating that this variation is fairly uniform throughout all the individuals of a flock, whether high producers in three years or low ones. It will also be observed that it does not make any difference whether the first year record is high or low. If the first year record is high as in the flock of 1908, 1909, and 1912, then the second and third year records are lower throughout than the first, while in the other three flocks with a low first year record, the second and often the third years are uniformly higher than the first.

It will also be observed that the three year averages of the highest tens in each flock are not influenced as much by high first year records as they are by the size of flocks. The smaller the flock the smaller will be the number of high producers.

It must be remembered in studying this table that this uniformity in variation was secured by arranging these fowls on the basis of the three year average. The difference in the results secured by selecting at the end of the first year will be noticed later in discussing Table No. XV.

Correlation between First, Second and Third Year Production.

Figures 6, 7, 8, 9 and 10 show the production of the first years of these flocks arranged in order, while the production of the same individual during the second and third years is shown on the same vertical line above or below. These figures show that in general the first year production is a poor guide to the production in later years. Instead, wide variations are observed in individual production during the three years, this variation occuring as frequently in the high layers as in the low ones.

---

[1] The 1912 flock is slightly over double the size of the others. To make comparisons between the flocks more accurate this flock has been considered in twenties instead of tens.

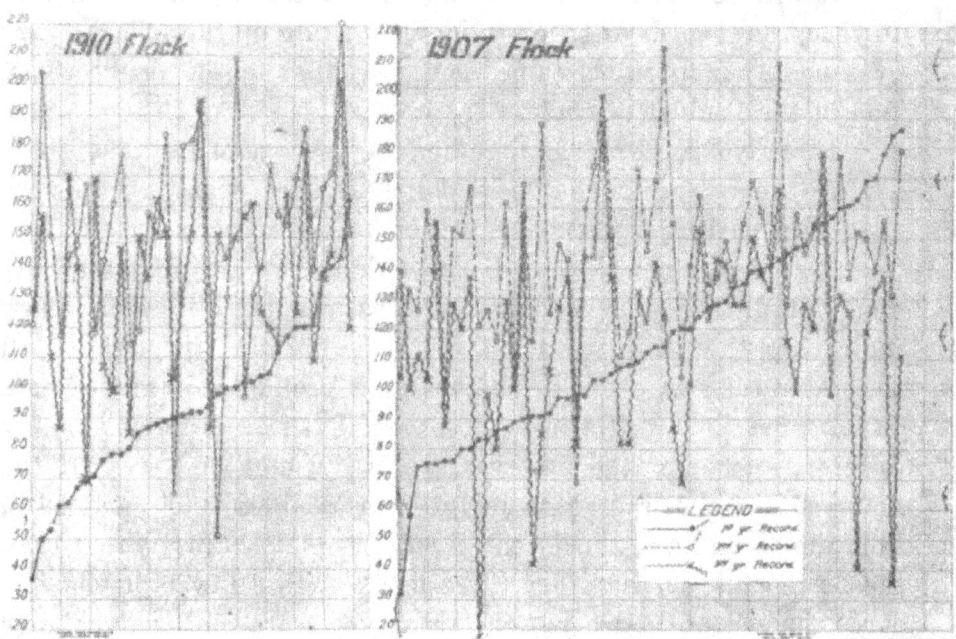

Fig. 6—A low first year flock average.

Fig. 7—A low or intermediate first year flock average.

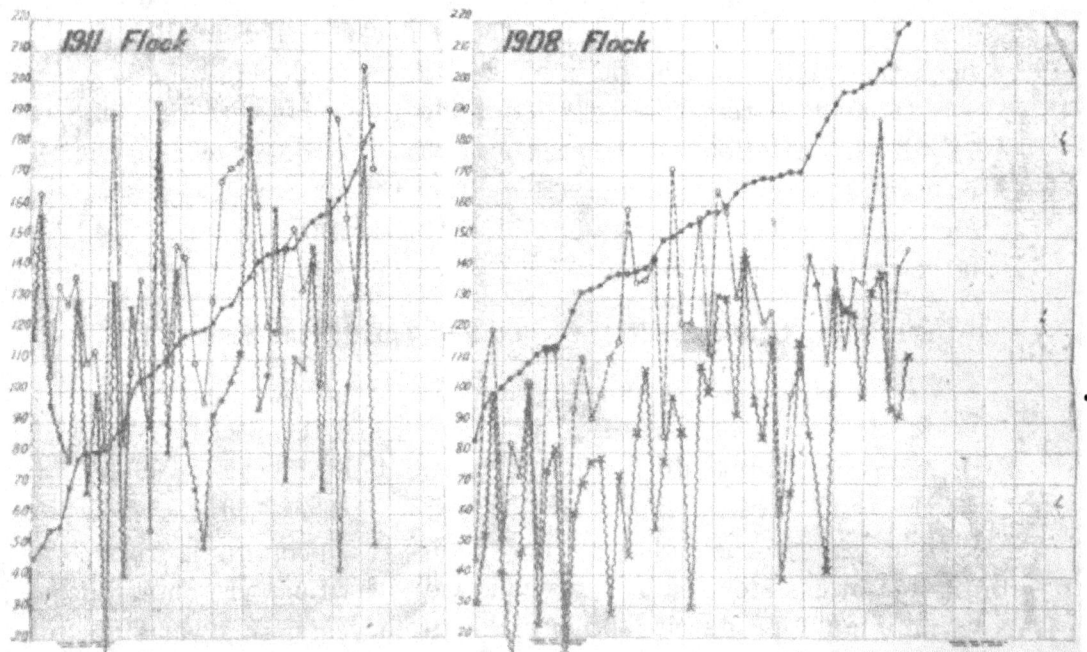

Fig. 8—A low or intermediate first year average.

Fig. 9—A high first year average.

Hens are ranked on first year record. Note the difference in second and third year records when the first year average is high or low.

ANNUAL EGG PRODUCTION 41

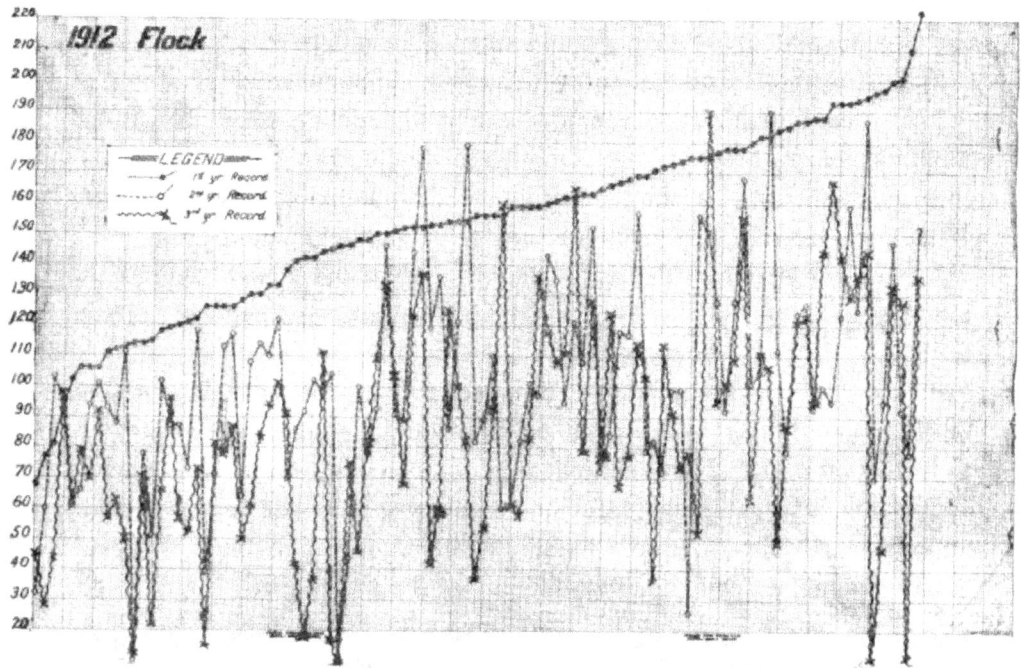

Fig. 10—Hens are ranked on first year record. The three records of each hen are given on the same vertical line. Note the high first year average followed by a lower second and third year record.

Dividing the flocks into those with high first years as in Figure 9 and 10, and those with low ones, Figures 6, 7 and 8, and it will be noticed that there is less variation in the second and third year records of the high flocks and a greater correlation with the first year. The reason that there is less variation in these records, in part at least, is that there is less production and therefore less chance for variation.

The 1911 flock, while intermediate in first year production, shows strikingly wide variation in individual records.

Table No. XIV gives the correlations calculated by the short method for the second and third years in relation to the first and the second and third in relation to each other. The flocks are arranged into two groups on high and low first year records and these averaged separately. The flocks are also arranged so that they fall in order on first year production, the highest first. The most striking thing about Table XIV is the almost perfect agreement between the height of the first year flock records and the correlation between first and second year. The highest record of 156 eggs shows a correlation of 63% which is quite high, while as the flock records fall the correlation falls correspondingly until the low record of 94 eggs shows only 17% correlation, which is very low in itself and not quite double the probable error.

The correlation between first and third year production does not fall as regularly. It is fairly high in the high flocks, but falls to almost nothing in the lower ones. It just reverses the order of the first and second year correlation in the low flocks.

The correlation between second and third year records is moderately high and fairly uniform throughout, averaging a little higher in the high flocks than in the low ones.

Taking all flocks together, it is seen that the correlation between second and third year production is considerably higher than the correlation between the first year and either of the others.

Nixon[20] gives correlation tables on 88 White Leghorns showing a first-second year correlation of 55% first-third year correlation of 15% and a second-third year

TABLE NO. XIV—CORRELATION BETWEEN FIRST, SECOND, AND THIRD YEAR PRODUCTION IN WHITE LEGHORNS.

Flock of	1st Year Record	High First Year Flocks Between 1st. and 2nd.	Between 1st. and 3rd.	Between 2nd. and 3rd.
1909	156	.6330±.0590	.5163±.0722	.6502±.0568
1908	153	.5631±.0658	.5533±.0668	.6804±.0518
1912	153	.3790±.0577	.3894±.0572	.5330±.0483
Average	154	.5250	.4863	.6212
Combination		.4856±.0368	.4365±.0390	.5527±.0335
		Low First Year Flocks.		
1911	119	.3366±.0958	.0905±.1071	.5018±.0808
1907	117	.2276±.0840	.1376±.0869	.5106±.0655
1910	94	.1698±.1077	.2536±.1038	.2862±.1018
Average	110	.2446	.1606	.4329
Combination		.1999±.0559	.0407±.0582	.4509±.0464
Average of all		.3848	.3234	.5270

correlation of 40%. These figures were based on a flock that averaged 92 eggs the first year, 97 the second year and 86 the third, or a three year average of 92, which is very low for Leghorns, and suggests that other factors probably interfered with production throughout the test.

She also gives the correlation between the years of laying and the three year total, but the record of the year, especially of a high one, is so large a factor of the three year total, that it often furnishes over one-half of the correlation shown.

Is First Year Production an Indication of the Future Production of a Hen?

Table No. XV shows the production of each flock arranged in tens according to the number of eggs they laid the first year. Examining the three year totals it will be observed that in every case the ten hens in the different flocks that laid the most the first year made the highest three year average and were, therefore, the highest producers for that length of time, and further that the three year averages decreased in the same relative order as the first year production, with only one or two exceptions, down to the lowest ten in each flock, so that, to that extent at least, the first year production is an accurate indication of the total three year production.

Dividing the flocks into two groups on first year production and taking the flock of 1907, where there were sufficient hens to give reliable distribution, as an example of the low first year group, and it is observed that there is a difference of 97 eggs to the hen between the lowest and the highest ten the first year. Taking the three year totals in the same way, we find a difference of 115 eggs, or subtracting the 97 it leaves a remainder of 18 eggs. This difference in two years is only 9 eggs a year variation between the highest layer and the lowest, after the first year. What is true of these hens is true of low first year flocks in general—that selecting the high first year layers would not materially raise the productivity of the flock during the next two years. In fact, in the 1907 flock three of the intermediate tens made higher records the next two years than the highest ten did. We can now see that the variation in the three year averages of the low first year flocks is practically all due to the variation in the first year records.

Taking the 1909 flock as an example of the high first year records and a very different condition is found. On an average there is about 100 eggs difference between the highest and lowest tens in the first year in all flocks regardless of production. In the low flocks there is only a little more than this difference in the three year totals, while in the high flocks, like that of 1909, this number is almost doubled and there is a difference of about 200 eggs between the highest tens and the lowest. This shows that in these flocks, selection on the basis of first year production would materially increase the flock average after the first year.

TABLE NO. XV—AVERAGE NO. OF EGGS LAID BY EACH TEN OF EACH FLOCK ARRANGED ACCORDING TO FIRST YEAR PRODUCTION.

Flock and Year	'08	'09	'10	'11	'12	'13	14	15	Total Production. Avg. 3 Yr.	3 Yr.
1907										
1st. 10	168	149	113	78	40	98	......	......	143	430
2nd. 10	143	156	131	79	86	89	44	62	143	429
3rd. 10	121	151	121	88	82	100	......	......	131	393
4th. 8	104	144	124	90	108	96	37	......	124	372
5th. 10	90	140	106	69	83	68	5	10	112	336
6th. 10	71	135	110	71	99	116	68	73	105	315
Flock Avg.	117	146	117	78	87	93	44	52	126	379
1908										
1st. 10		202	137	109	117	85	73	87	149	448
2nd. 10		171	120	91	111	99	85	75	127	382
3rd. 9		153	137	91	96	109	57	49	127	381
4th. 10		134	108	65	53	39	52	22	102	306
5th. 10		104	85	57	66	99	34	28	82	246
Flock Avg.		153	117	82	90	89	58	58	118	353
1909										
1st. 10			200	140	132	125	70	48	157	472
2nd. 10			181	119	125	112	77	50	142	426
3rd. 8			160	114	111	91	71	2	128	385
4th. 10			139	114	99	90	73	97	117	352
5th. 10			101	71	96	92	45	51	89	268
Flock Avg.			156	111	113	103	65	47	127	380
1910										
1st. 10				128	169	141	26	25	146	438
2nd. 10				96	144	145	95	65	127	387
3rd. 7				85	138	133	76	54	119	357
4th. 10				62	147	124	57	34	111	334
Flock Avg.				94	151	136	76	53	127	381
1911										
1st. 10					163	157	110	98	143	430
2nd. 10					134	146	107	74	129	388
3rd. 9					107	120	101	77	109	328
4th. 10					68	131	98	59	99	296
Flock Avg.					119	139	104	75	120	361
1912										
1st. 20						193	118	103	138	414
2nd. 20						171	110	101	128	383
3rd. 20						156	110	93	120	359
4th. 20						139	93	69	100	299
5th. 20						107	77	57	80	241
Flock Avg.						153	101	85	113	339

In general then these studies indicate that where a flock has made a high first year record, selecting the high producers will increase the later production of the flock and will at the same time give a definite basis for determining the productivity of the individuals, but on the other hand in a flock with a low first year record the highest producers are likely to come mainly from the intermediate tens and selection cannot be made until later.

It will also be seen that even though the high producers had been selected from a high flock their record for the next two years would not have been as high as that of a flock with a low first year record. For example, the highest ten of the highest first year flock (1909) laid an average of 272 eggs in the second and third years, while the total flock average of the 1910 flock (the lowest flock) was 281 and the highest ten laid 310 eggs. The three intermediate tens of the low 1907 flock laid more eggs on an average than the highest ten of the 1909 flock. In fact the lowest tens of the low flocks averaged 8 more eggs in the second and third years than the highest tens of the high flocks did, and as shown in Figure 2, the low flocks as a whole averaged higher three year totals than did the high flocks.

In this connection it is well to remember that it was shown in a previous publication[1] that the ten highest second year hens gave a higher three year average than the ten highest first year hens.

TABLE NO. XVI—AVERAGE PRODUCTION OF EACH TEN OF EACH FLOCK ARRANGED ACCORDING TO FIRST YEAR RECORDS (CORNELL FLOCKS).

Flock A.	'09	'10	'11	3 Yr. Total	Flock B.	'09	'10	'11	3 Yr. Total
1st 10	195	151	134	480		201	167	157	526
2nd 10	168	138	127	432		148	130	115	393
3rd 8	150	139	121	410		124	112	97	333
4th 10	102	112	109	324		89	85	64	238
Flock Avg.	154	135	123	411		142	124	109	375

Rice[28] gives the only other records of an unselected flock in which three years of individual records are given. These flocks were both of the same year, but for some reason have been treated separately. Their production by tens arranged on first year record is given in Table XVI. They are both high first year flocks with lower second and third year records, but beyond that they are quite different. Flock A did not have a single extremely high layer nor a very low one, but the averages were all high throughout all three years, not a single average falling below 100, and the flock average of three years of 137 was exceptionally high. Flock B on the other hand had three extremely high layers and the rest of the fowls dropped far below, running down to a very low layer at the bottom. This flock shows the extreme of influence of first year production, there being a difference of 288 eggs between the highest and lowest tens. If all flocks were like this it would be easy to select high producers.

A comparison of the same kind of a record from a selected flock will show the difference in variation in the different tens. Opperman and Waite[21] selected the 60 best layers out of a flock of 240 and kept them three years with results as shown in Table XVII.

TABLE NO. XVII—AVERAGE PRODUCTION OF EACH TEN OF FLOCK ARRANGED ON FIRST YEAR PRODUCTION. (MD. FLOCK.)					
Flock of			'08	'09	'10	
1907	1st. 10	204	161	119	484
	2nd. 10	183	144	111	439
	3rd. 10	173	169	132	473
	4th. 10	164	148	124	436
	5th. 10	157	132	94	384
	6th. 10	147	141	111	398
Average		171	149	115	
Average Orig. flock		127	114		
Average remainder		112	103		

It will be observed that there was a difference of only 57 eggs in the first year between the highest and lowest tens instead of 100, as shown by the unselected flocks and that the three year totals only differed by 86. Although this is an extremely high first year average, the flock from which they were taken only averaged 127 so that these should represent the highest tens of a medium flock. In second and third year production they are somewhat intermediate, with the exception of the third ten, which with its record of 301 for these years equals the low flocks. The 127 first year record was this flock's highest record evidently, as the second year record was only 114, so this should be classed as a high first year flock.

The Difference between the Highest and Lowest Producers in the Flocks.

While the flocks average practically the same production in three years, they are of course made up of individuals which vary all the way between very high and very low production each year, and in which the three year totals vary by several hundred eggs. Table XIII shows the averages of the tens based on the total three year production. On studying this table it will be noticed that there is very little difference in totals between the highest tens of the different flocks regardless of the difference in first year production.

Dividing them into groups on high and low first year production we get the following averages of the highest tens:[1]

	1st. Yr.	2nd. Yr.	3rd. Yr.	3 Yr. Total
Highest tens of high first year flocks average	188	152	135	476
Highest tens of low first year flocks average	139	174	151	465
All flocks average	163	163	143	470

showing a difference of only 11 eggs in three years in favor of the high flock.

Studying the lowest tens in the same way and their total production is found to be fairly uniform, but upon tabulating them it is found that the totals are reversed, the low first year flocks

	1st. Yr.	2nd. Yr.	3rd. Yr.	3 Yr. Total
Lowest tens of high first year flocks average	114	66	50	229
Lowest tens of low first year flocks average	81	108	80	273
All flocks average	97	87	65	251

showing the highest total production and a much greater difference than was shown by the high flocks.

Taking the averages of these two tabulations and comparing them gives the difference between the highest and lowest tens of all flocks, showing little variation between the years, but slightly

---

[1] The highest and lowest twenties of the 1912 flock were taken instead of the tens as this flock was nearly double the size of the others and this makes about the same proportion compared.

	1st. Yr.	2nd. Yr.	3rd. Yr.	3 Yr. Total
Highest tens average	163	163	143	470
Lowest tens average	95	87	65	251
Difference	68	76	78	219

increasing with age, with an average difference of 73 eggs per year or 219 eggs per hen in three years.

Taking the high first year flocks alone gives an average difference of 82 eggs per year, or 245 in three years while three low flocks average a difference of 64 per year or 189 for the three years.

In Table No. XV the flocks are arranged on first year records. Comparing the highest and lowest tens in the same way,

	1st. Yr.	2nd. Yr.	3rd. Yr.	3 Yr. Total
Highest tens of high first year flocks	198	132	115	445
Highest tens of low first year flocks	153	158	121	433
All flocks average	176	145	118	439

a higher first year average and lower production afterwards is obtained, including a lower total than the high tens on the previous arrangement. They show only 12 eggs difference in the total between the high flocks and the low ones. The low tens from this table show a lower first year with higher production

	1st. Yr.	2nd. Yr.	3rd. Yr.	3 Yr. Total
Lowest tens of high first year flocks	104	78	70	252
Lowest tens of low first year flocks	67	138	111	315
Low tens all flocks	86	108	91	284

thereafter, including the total, than on the previous arrangement. They also show a higher total production for the low tens of the low flocks, in this case amounting to 63 eggs.

Comparing the averages of all high tens and low tens on first

ANNUAL EGG PRODUCTION

year production, shows a higher first year difference but much less

	1st. Yr.	2nd. Yr.	3rd. Yr.	3 Yr. Total
All high tens	176	145	118	439
All low tens	86	108	91	284
Difference between high and low tens	90	37	27	155

difference decreasing with the age in the following years and a total difference of 155 as against 219 on the other arrangement.

Taking the difference in the high and low flocks separately gives the following:

	1st. Yr.	2nd. Yr.	3rd. Yr.	3 Yr. Total	After 1st. Yr.
Difference between high and low tens, high flocks	94	54	45	193	99
Difference between high and low tens, low flocks	86	20	10	118	30

This shows that when selecting on the first year records there is very little difference in later production between the highest layers and the lowest ones of the low flocks.

TABLE NO. XVIII—AVERAGE PRODUCTION OF HIGHEST AND LOWEST TENS AND TWENTIES OF ALL FLOCKS.

	1st. Year	2nd. Year	3rd. Year	Average 3 Years	Total 3 Years
Arranged on 3 Year Average.					
Highest Tens	163	163	143	157	469
Next Highest Tens	145	141	119	135	405
Next Lowest Tens	120	115	93	110	328
Lowest Tens	98	88	67	84	253
Highest Twenties	154	152	131	146	437
Lowest Twenties	109	102	80	97	291
Arranged on First Year Production.					
Highest Tens	176	145	118	146	439
Next Highest Tens	149	132	117	133	399
Next Lowest Tens	116	119	95	110	330
Lowest Tens	85	108	90	94	283
Highest Twenties	162	139	117	140	419
Lowest Twenties	100	114	93	102	307

It must also be remembered in this connection that the intermediate tens of the low flocks have often produced more in the later years than the highest tens and that the low flocks have produced more in three years than the high ones.

The average production of the highest and lowest twenties of all flocks under both arrangements is given in Table No. XVIII. This corresponds closely with the production of the upper and lower halves of the flocks, as it includes all of two flocks and all but the fractional tens of three more.

The striking thing about this table, is the remarkably close agreement in the variation from the mean of the opposite tens under both systems of arrangement. In practically every case the average of the extreme or intermediate tens will give within one or two eggs of the mean or general average of the flock indicating that the variation in the individuals of the flocks is absolutely normal and uniform throughout.

The Year in which the Highest Record was made.

In the high first year flocks a great majority of the hens make their highest record the first year. Of those that do not, some make it the second, but are equally likely to wait until the third or fourth year and three hens have made their highest record in the fifth year, as is shown in Table No. XIX.

TABLE NO. XIX—THE YEAR IN WHICH THE HIGHEST RECORD WAS MADE AND THE TOTAL NUMBER OF RECORDS HIGHER THAN THE FIRST YEAR.

	Hens with Records for Three or More Years							Hens with Only 2 Years Record	
	No. of Hens Making Highest Record							No. Making Highest Record	
Flock of	1st. Yr.	2nd. Yr.	3rd. Yr.	4th. Yr.	5th. Yr.	Total No. of Records above 1st. Year.	No. of Hens.	1st. Yr.	2nd. Yr.
1907	13	41	4	0	0	96	58	15	18
1908	38	7	0	1	3	14	49	39	10
1909	39	4	3	2	0	12	48	24	1
1910	1	24	12	0	0	83	37	8	55
1911	12	24	3	0	......	49	39	8	6
1912	92	4	4	......	......	8	100	35	2
1913	......	......	......	......	......	......	......	34	63
Total	195	104	26	3	3	262	331	163	155
		136				From 3 Yr. Records		195	104
						Grand Total		358	259

In the low first year flocks the majority of the hens make their highest record in the second year, some waiting until the third, but none later in our records. Of these hens with low first year records, quite a number have continued to make higher records than the first year for several years, one hen continuing for five years to exceed the first year record.

Taking the six flocks with three year records and 195 hens made the highest record the first year, 104 the second year, 26 the third year, 3 the fourth and 3 the fifth year, or 136 in all after the first year. The same hens made a total of 262 records higher than the first year ones.

Hens that lived only two years together with the 1913 flock which has only two years record to date, gave 163 first year records highest and 155 second year, which added to the former figures gives a grand total of 358 highest first year records, 259 second year records, and 32 after the second year. If we add to the 259 second year records the 32 later ones and the 126 records higher than the first year, but not the highest, we would have a total of 417 records higher than the first year against 358 highest first year records. In these flocks over one-half of the highest records have been produced after the first year.

In flocks A and B (Cornell)[28] 40 made the highest record the first year, 15 the second and 11 the third, and there were 15 other records higher than the first year, or 41 records in all higher than the first year ones. The Maryland[21] flock was selected on high first year records so these would of course in all cases be the highest. There must have been a considerable number of higher second year records in the 180 remainder of the flock, as the first year average was only 112 and the second year 103. Miss Nixon's[20] flock of 88 hens made a higher average the second year record as well as a third year average only a few eggs below the first, so that a majority of these hens must have made higher records after the first year.

The Lengh of Life of a Hen.

What the average life of a Leghorn hen would be if allowed to live it out, is still unknown. The data in Table No. 1 is practically the only material at hand and unfortunately in the early years of the experiment some culling was done and there have always been a few lost and a few stolen each year. None of the flocks have lost all their hens yet, so it is too early to do more than estimate. From the figures in the table the average life is

4½ years from hatching time. Most of them live and produce in their last incomplete year—not counted in the table—up until hot weather, which would add one-half year. Another year at least should be added to make up for culls and losses, so that the total life as estimated at this time would be six years, or one-half year development and five and a half productive years.

The Total Production of a Hen.

The total possible production of a hen depends partly upon the length of life, and as some of the members of the oldest flocks are still living and producing, it is too early to do more than estimate total production.

Table No. XX shows the production by hundreds of all six flocks to date. Of the three older flocks with only 28 hens living, the record is fairly complete, although some of these will no doubt move up a few hundreds yet before death. Of the younger flocks with 151 living out of 176 that finished the third year it is too early to even speculate.

TABLE NO. XX—THE TOTAL PRODUCTION BY HUNDREDS OF THE MEMBERS OF EACH FLOCK TO NOV. 1, 1915.

Flock of	No. of hens which finish 3 years	No. of hens living Nov. 1915	No. of hens that have laid				
			over 500	over 600	over 700	over 800	over 900
1907	58	4	27	17	9	2	1
1908	49	10	23	14	8	2	0
1909	48	14	27	10	3	1	1
1910	37	18	12	5	1	0	0
1911	39	33	11	3	0	0	0
1912	100	100	4	0	0	0	0
Total	331	179	104	49	21	5	2

One-half of the hens of the first three flocks that lived three years had laid before November 1st, 1915, between 500 and 600 eggs, over one-fourth of them between 600 and 700, over one-eighth 700 to 800, one thirtieth 800 to 900 and two between 900 and 1000.

One-third of the hens of the fourth flock were above 500 and one-seventh above 600, while one had reached 700. In all, 21 hens have passed the 700 mark, 5 the 800 and 2 the 900, with Queen Utahna still in the lead with 916 eggs to November 1915, 948 at close of 1916 laying.

Handrick(9a) gives the monthly record of a hen that laid 1034 eggs in eight years and was then killed. The successive yearly records were as follows:

105-163-138-159-160-133-111-65 total 1034 eggs.

Table No. XXI shows the annual production of all hens in our flocks that have made records above 700, arranged to show the years in which the records were made, as well as the comparative sequence of production. One-half of these hens started in unfavorable years and all but two made higher records in the second year. The other half started in favorable years and all but three made their highest record the first year. There seems to be no relation between the first year production and the total of these hens.

TABLE NO. XXI—THE YEARLY PRODUCTION OF HENS LAYING OVER 700 EGGS.

Hen No.	'08	'09	'10	'11	'12	'13	'14	'15	1st	2nd	3rd	4th	5th	6th	7th	8th	3 Yr. Avg.	Total to Date
200	103	197	188	72	108	103			103	197	188	72	108	103			163	771
204	155	154	120	112	106	91	37		155	154	120	112	106	91	37		143	775
213L	148	145	128	103	107	86	26	18	148	145	128	103	107	86	26	18	140	761
279L	141	144	133	105	116	95	61	106	141	144	133	105	116	95	61	106	139	(901)*
211L	105	151	137	104	104	88	37		105	151	137	104	104	88	37		131	726
352	127	135	123	87	101	100	(27)		127	135	123	87	101	100	(27)		128	700
274	75	139	155	47	89	102	50	73	75	139	155	47	89	102	50	73	123	(730)
325L	91	188	85	86	147	144	(22)		91	188	85	86	147	144	(22)		121	763
278L	80	151	120	87	144	131	86	(12)	80	151	120	87	144	131	86	(12)	117	811
743		197	137	125	113	103	73	110	197	137	125	113	103	73	110		153	(858)
555		158	165	131	130	110	(47)		158	165	131	130	110	(47)			151	741
841		169	126	117	101	115	54	52	169	126	117	101	115	54	52		137	(734)
734		216	140	92	134	125	(48)		216	140	92	134	125	(48)			149	755
729		176	144	86	108	130	115	97	176	144	86	108	130	115	97		135	(856)
766		199	135	98	106	125	90	(0)	199	135	98	106	125	90			144	753
551		206	101	94	105	90	92	90	206	101	94	105	90	92	90		134	(778)
562		150	172	98	117	137	99	(0)	150	172	98	117	137	99			140	773
96U			195	193	138	161	129	97	195	193	138	161	129	97			175	(913)
106U			203	149	149	152	84	(29)	203	149	149	152	84	(29)			167	766
125U			158	175	150	130	108	(0)	158	175	150	130	108				161	721
43				100	209	150	147	136	100	209	150	147	136				153	(742)
Average									150	155	125	110	115	98	65	66	143	778

Several interesting points are noted in this table: out of the 20 odd hens that have laid 200 eggs or over in a year in these flocks, only three appear here. Five hens have passed the 800 mark, one of these has the highest three year average and one the lowest. One hen has six records above that of the first year and several have four. There have been records above 100 eggs in every year of laying including the eighth and last.

The average production of these hens for the first three years was 150, 155 and 125 with a three year average of 143. This average is three eggs below the average of the highest 20 hens in each flock. This, and the low first year records, indicate that it

*Totals in brackets are of hens still living.

is not as much early fecundity as consistent laying that tends to high totals. Of the 18 hens from the first two flocks, 4 came from the highest tens, 7 from the second tens, 4 from the third tens, 2 from the fourth, 1 from the fifth and none from the lowest ten in either flock as arranged on the three year average.

Comparative Production of the Different Breeds.

All breeding work at this station in recent years has been done with S. C. White Leghorns. Believing that the results obtained should be checked with one of the general purpose breeds, a small flock of White Wyandottes was recently introduced, but no comparative records have as yet been obtained.

The White Leghorns appear to be the most popular and productive egg laying breed at the present time. Purvis[26] gives a table showing the comparative number and production of the most popular breeds in the 5 American laying contests of 1915, part of which has been included in the Table No. XXII. To these have been added the records of two Australian and two British contests, representing the other countries in which contests are held, only those breeds with 50 or more entries in some one contest are included and only the highest six breeds from the American contests. Only three breeds are represented in Australia by any numbers and only two in the British contests. The Buff varieties are more popular in the British contests but none of them had as many as 50 entries.

TABLE NO. XXII—NUMBER OF INDIVIDUALS AND AVERAGE PRODUCTION OF LEADING BREEDS IN THE 1915 LAYING CONTESTS.

	American Contest 1915		Australian Contests 1915		British Contests		Total No.	%
	No.	Avg.	No.	Avg.	No.	Avg.		
White Leghorn	891	163	742	193	134	201	1767	57
White Wyandotte	205	156	12		188	185	405	13
R. I. Reds	378	158	6		16		400	13
Barred Rocks	161	152	12				173	6
W. Rock	110	148	18				128	4
W. Orpington	80	114	24				104	3
Black Orpington			96	158			96	3
S. Wyandotte			48	163			48	1
Totals	1825		958		338		3121	100
					Total White breeds		2404	77

From Table No. XXII it is seen that the White Leghorns were represented in these contests by more individuals than all other breeds combined, and by over four times as many as their nearest competitor. No other breed was represented in all contests in sufficient numbers to warrant averaging production, but the Leghorn gave the highest production in every case, and as shown in the discussion of laying contests have almost always led in the past.

Another interesting thing brought out in Table XXII is the great preponderance of white birds, 77% of the individuals shown being white and 16% red or black, while only 7% are of the parti-colored varieties. It appears from this that the utility bird of the future will be a solid colored bird preferably white.

SUMMARY AND CONCLUSIONS.

The study of eight years records of a flock of White Leghorn hens combined with a similar study of the production of each successive flock of the unselected descendants together with a review of the literature furnishes the basis for the following summary and conclusions.

The average production of the successive years of the same flock under our conditions has been as follows:

Year	1st.	2nd.	3rd.	4th.	5th.	6th.	7th.	8th.
No. of eggs	124	119	106	82	74	66	51	52

The average first year production of other unselected flocks kept under favorable conditions has been about 128 eggs.

The first year average of a flock may vary between 95 and 160 eggs as influenced by maturity and environment.

The second year average of a flock may vary between 100 and 145 eggs as influenced by environment and previous production.

A flock that makes a low first year record as the result of unfavorable conditions will make a higher record the second year.

The average of the highest year of each of our flocks is 139 eggs for all hens, and 150 eggs for those hens that finished three years of laying.

The total productions of our flocks in three years has been very constant, regardless of whether the first year's production was very high, very low or intermediate, in fact:

The three year total of the low first year flocks has on the average been higher than that of the high first year flocks.

The total production for the fourth, fifth and sixth years has been fairly constant, varying slightly inversely with the first year production but apparently not affected by the three year total.

The average first year production of selected flocks of White Leghorns (egg laying contests) is about 162 eggs for American contests and 190 for Australian.

The average first year production of selected flocks of the best of general purpose breeds (egg laying contests) is about 153 eggs for the American contests and 165 eggs for the Australian.

The highest ten hens in each of our six flocks, averaged in the first year 176 eggs, which is higher than the American contest records.

The highest ten hens in the highest years of each of our six flocks averaged 194 eggs, which is higher than the Australian contests, showing that their selection has not been as rigid as to equal the highest fifth of an average flock.

The first year record of a flock is largely a matter of maturity and may be increased by selection for maturity.

There is little difference in our flocks in average productivity between the first and second years and the third year drops between one and two dozen on the average, the three year's records being 132, 127 and 106 for those that lived three years or more.

The three year's records for other nonselected flocks of White Leghorns average 127, 116 and 110.

No records of nonselected flocks of general purpose breeds extending over two or more years have been found.

The average production of all selected flocks whose records were available for the three years was:

Leghorns	180	146	119
General purpose breeds	177	116	93

The average three year production for the highest ten hens in each of our flocks of White Leghorns was 176, 146 and 118, indicating about the same degree of selection as in egg laying contests.

The correlation between first and second year production varied directly with the production of the flock, ranging from almost nothing in the lowest flock up to 63% in the extremely high ones.

The correlation between second and third years is considerably higher than between the first year and either of the other years.

Variations in egg production are largely due to seasonal (climatic) influences, the first years record being affected by the environment from hatching time on.

The three yearly records of an individual vary more widely on the average than the flock records. If one record is far above the flock average, another is fairly certain to be far below.

Where a flock makes a low first year record, there is little difference in the future production between the highest and lowest individuals, but that all production will be high.

Where a flock makes a high first year record the high producers will on the average continue to be much higher producers than the low ones, but that all later production will be low.

The highest tens of the high flocks on first year production laid 99 eggs more in the next two years than the low tens.

The highest tens of the low flocks on first year production laid only 30 eggs more in the next two years than the lowest tens.

The intermediate tens of the low flocks often laid more in the later years than the high tens.

In these flocks 40% of the hens have made their highest record after the first year and extending in a few cases to the fifth year.

The average life of these fowls so far as can be determined at present is between 5 and 6 years.

The total possible production is probably above 1000 eggs. The upper one-half of the first two flocks have produced an average of over 600 eggs each.

The average production of the entire 1907 and 1908 flocks to November 1st, 1915, was 503 and 494 eggs respectively and 14 hens were still living at that time.

One-half of the hens that have produced over 700 eggs have had low first year records. They have come from all tens except the lowest and more from the second than from the first.

The White Leghorn is at present the most productive breed of fowls and by far the most popular for egg production, so that studies based upon this breed are the most valuable to egg producers.

CONCLUSIONS.

The production of unselected White Leghorns varies widely in different years as influenced by the environment but from all available records averages about 130 for the first year, 120 for the second and less than 110 for the third, drops to about 85 in the fourth and falls about 10 eggs a year after this up to the eighth year. Selected flocks have averaged 160 in America and 190 in Australia. The American record corresponds closely to the average of the upper one-half of the unselected flocks and indicates that the selection has been able to eliminate the lower half.

The first year production of a flock of White Leghorns is no indication of their total production, if the first year is high the second will be low, if the first is low the second will be high, but the total production in three years will in all cases be about the same.

If the first year record of a flock is high, selection of the high layers will materially improve the later production of the flock. If the first record is low there will be little value in selection as even the lowest producer will make a second year record above the general average. The three year average is in all cases a much more reliable indication of productivity.

The average life of a White Leghorn appears to be about 6 years. The average production of the fourth year is equal to the average production given for the United States. The average total production is above 500 eggs and the maximum possible production above 1000.

The White Leghorn is the most important egg producing breed at the present time, over one half of all contest entries are Leghorns. Their average production has been decidedly above the average of the general purpose breeds. Three-fourths of all contest entries have been white.

BIBLIOGRAPHY

(1) Ball, Turpin, and Alder, 1914. "A Study in Annual Egg Production. Utah Exp. Sta. Bul. No. 135.

(2) Dryden, Jas. 1905. "Poultry Experiments." Utah Sta. Bul. No. 51, 60 and 92.

(3) 1907. "Poultry Experiments." Utah Sta, Bul. No. 102.

(4) Dunnicliff, A. A. 1913. "Ten Years' Egg-Laying Tests at Hawkesbury Agr. Col. and Exp. Farm. Dept. of Agr. New South Wales. Farmers' Bul. No. 66.

(5) 1915. "Thirteenth Year's Results of the Hawkesbury Agr. Col. Laying Contest." Dept. of Agr. New South Wales. Farmers' Bul. No. 103.

(6) Gilbert, A. G. "Report of the Poultry Manager." Canada Exp. Farms Reports 1895 to 1906 inclusive.

(7) Gowell, G. M. 1900. "Poultry Exp. in 1899." Maine Agr. Exp. Sta. Bul. No. 64.

(8) 1902. "Poultry Exp. in 1900 and 1901." Main Agr. Exp. Sta. Bul. No. 79.

(9) 1903. "Poultry Exp. in 1902." Main Agr. Exp. Sta. Bul. No. 97.

(9a) Handrik, Mrs. 1910. Ztschr, Landw. Kammer Schlesien 14, No. 15 pp. 451-453.

(10) Irish. Dept. of Agr. 1908. "A Survey of 125 Flocks Representing over 5,000 fowls." The Journal of the Board of Agr. Vol. XVI No. 7, p. 585.

(11) Jacobs, W. S. "Farm Poultry." 1908. Arkansas Agr. Exp. Sta. Bul. No. 99.

(12) Kirkpatrick, Wm. F. Nov. 8, 1912. "International Egg Laying Competition." Storrs Conn. Press Bul.

(13) Nov. 5, 1913. "The Second International Egg Laying Contest." Press Bulletin. Storrs, Conn.

(14) Card, L. E. 1915. "Third Annual International Egg Laying Contest." Conn. Storrs. Agr. Exp. Sta. Bul. No. 82.

(15) Lewis H. R. Reports of the Agr. Exp. Sta. of New Jersey for 1911, and 1913.

(16) Laurie, D. F. "Official Reports of Egg Laying Competitions." 1912-13-14-15. Dept. of Agr. South Australia.

(17) Lloyd, W. A. and Elser, W. L., 1911. "A Cooperative Investigation of the Profitableness of Poultry under Farm Conditions. Ohio Agr. Exp. Sta. Cir. No. 118.

(18) Linfield, F. B. 1906 Annual Rpt. Mont. Agr. Exp. Sta.

(19) Murphy, Miss L. Rpts. of the First, Second and Third Irish Egg-Laying competitions. 1912 to 1915.

(19a) Nelson, Julius "Reports of the Agr. Exp. Sta. New Jersey for 1905 and 1906.

(20) Nixon, Clara, 1910, "A Study of the First, Second and Third Year Egg Production of White Leghorn Hens." Ann. Rpt. Amer. Breeders' Ass'n. Vol. VII p 279-288.

(21) Opperman, C. L. and Waite, R. H. 1911. Some Exp. with Poultry." Maryland Agr. Exp. Sta. Bul. No. 157.

(22) Pearl R. and Surface, F. M. 1909. "A Biometrical Study of Egg Production in the Domestic Fowl."
1. Variation in Annual Egg Production U. S. D. A. Bureau of An. Ind. Bul. No. 110, Part 1 and 2.

(23) Pearl, R. 1911. "Breeding Poultry for Egg Production." Maine Agr. Exp. Sta Bul. No. 192.

(24) 1915. "Seventeen Years Selection of a Character." The Amer. Naturalist. Vol. XLIX. P. 586-595.

(25) Philips, A. G. 1915. "Poultry Investigations: The Value of Meat Scrap, Fish Scrap, and Skim Milk in Rations for Laying Pullets," Indiana Agr. Exp. Sta. Bul. No. 182.
(26) Purvis, Miller; "The Laying Limit of Hens." Breeders' Gazette, Vol. LXVIII. Dec. 2, 1915, p. 1005.
(27) Quisenberry, T. E. Reports of the Mo. National and International Egg Laying Contests for 1912, 1913 and 1914.
(28) Rice, Jas. E. 1913. "Observations on the Distribution of Daily Egg Production." The Cornell Countryman. Vol 10, No. 5, No. 6, and No. 7.
(29) 1915. "A Definite Program in Breeding for Egg Production." Penn Dept. of Agr. Bul. No. 267.
(30) 1908. "The Moulting of Fowls." N. Y. Cornell Exp. Sta. Bul. No. 258 and No. 318, 1912. Constitutional Vigor.
(31) Sherwood, R. M. and Buss, W. J. 1913. "Experiments with Poultry." Ohio Agr. Exp. Sta. Bul. No. 262.
(32) Stewart, J. H. and Atwood, H. 1899. "Poultry Experiments. W. Va. Agr. Exp. Sta. Bul's. No. 60 (1899), No. 88 (1903), No. 102 (1906), and No. 115 (1908).
(33) Taylor, H. C. 1911. "The Prices of Farm Products." Wisconsin Agr. Exp. Sta. Bul. No. 209.
(34) Terry, J. R. 1913. "Report of the Second International Egg-Laying Contest. Dept. of Agr. British Columbia.
(35) 1914. "Report of the Third International Egg-Laying Contest. Dept. of Agr. British Columbia.
(36) Thompson, D. S. Rpts. of the 8th to the 13th. Egg-Laying Competitions of the Hawkesbury Agr. Col. and Exp. Farms, New South Wales. Dept of Agr. Farmers' Bul. No. 44, 48, 57, 70, and 89 from 1910 to 1914 inclusive.
(37) Utility Poultry Club's Twelve Months' Laying Competitions Reported in Jour. of the Board of Agr.
(38) Wheeler, Wm. P. 1896. "Feeding Exp. with Laying Hens." New York, Geneva Agr. Exp. Sta. Bul. No. 106.
(39) 1896. Feeding Experiments with Laying Hens." Fifteenth Ann. Rpt. New York Geneva.
(40) 1891. "Feeding Exp. with Laying Hens." N. Y. Geneva Exp. Sta. Bul. No. 29. New Series.

www.ingramcontent.com/pod-product-compliance
Lightning Source LLC
Chambersburg PA
CBHW080000230526
45470CB00008B/2811